Reducing Soil Water Evaporation with Tillage and Straw Mulching

Reducing Soil Water Evaporation with Tillage and Straw Mulching

S. K. Jalota and S. S. Prihar

IOWA STATE UNIVERSITY PRESS / AMES

S. K. Jalota is a senior soil physicist at Punjab Agricultural University, Ludhiana, India.

S. S. Prihar is a former professor of soil physics at Punjab Agricultural University, Ludhiana, India.

The authors thank the following for permission to reprint the following figures and tables:

American Society of Agricultural Engineers: Table 5.2
Soil Technology: Figure 6.2; Table 6.2
Soil Use and Management: Figures 2.13, 3.7, 6.1; Tables 3.5, 3.6, 3.7, 5.3, 6.1
Australian Journal of Soil Research: Figures 2.7, 2.8, 2.10, 3.5, 4.9; and Table 4.4
Advances in Soil Science: Figures 2.11, 4.1, 4.2, 4.10; Table 4.2
Soil Science Society of America: Figures 4.6, 4.7, 4.8, 6.5; Tables 4.3, 5.1, 5.4, 6.4, 6.5

© 1998 Iowa State University Press, 2121 South State Avenue, Ames, Iowa 50014

Orders: 1-800-862-6657
Fax: 1-515-292-3348

♾ Printed on acid-free paper in the United States of America

First edition, 1998

Library of Congress Cataloging–in–Publication Data

Jalota, S. K.
    Reducing soil water evaporation with tillage and straw mulching / S.K. Jalota and S.S. Prihar. — 1st ed.
        p.   cm.
    Includes bibliographical references (p.   ) and index.
    ISBN 0-8138-2857-0
    1. Soil moisture conservation.   2. Evaporation.   3. Soil management.   I. Prihar, S. S. (Sohan Singh).   II. Title.
    S594.J35   1998
    631.5'86—dc21   97–48257

The last digit is the print number: 9   8   7   6   5   4   3   2   1

Contents

Preface

According to reports by experts, the availability of freshwater supplies is likely to be a major limiting factor in meeting the future agricultural needs of the growing world population. Conservation and efficient use of water hold the key to alleviating the constraint. Direct evaporation of water from the soil surface constitutes a large fraction of the total water loss not only from bare soil but also from cropped fields and is an unproductive water loss. Certain percipient agronomists, soil scientists, and meteorologists have realized that this loss must be reduced to make more water available for crops. Using theoretical analyses and experimental data from controlled laboratory and field experiments, they have furthered our understanding of the process of evaporation from soil and devised means to reduce it. Thanks to the efforts of these investigators, the evaporation process and its practical implications are much better understood today than they were a few decades ago. Major credit is owed to the early workers, who laid the groundwork and showed the path to be followed by others.

Although much theoretical and empirical information on the subject has accumulated, it lies scattered in various scientific journals, reports, and bulletins. This book compiles this information in one place and presents it systematically. The book will be useful to researchers, teachers, students, and planners. It will help in the selection of suitable water conservation management practices for various soil and climatic conditions.

We sincerely thank our colleagues in the Department of Soils, Punjab Agricultural University, Ludhiana, for their help and cooperation. Our publishers, especially Gretchen Van Houten, also deserve our thanks for their enthusiasm for the project, their cooperation, and their high production standards.

We humbly solicit comments and suggestions from our readers for further improvement.

1

Introduction

Water is essential for agricultural production and other life processes on the planet Earth. But its continued availability in suitable quality and in amounts sufficient for producing enough food, fiber, fuel, and shelter material for the projected population of the world is uncertain. Demand for nonagricultural uses of water—namely, domestic, industrial, and recreational—is increasing with the increase in population and living standards. Therefore, less and less water will be available for crop production in the future. A recent report by the Department of Water and Environmental Studies of the University of Linköping, Sweden (Falkenmark, Lundqvist, and Widstrand, 1990), has stressed that water scarcity will be the ultimate constraint for Third World development. Thus, we must use water with utmost efficiency in all the diverse areas, and particularly in crop production, which accounts for the major share.

It is well established that dry matter as well as grain production is directly related to water transpired by the crop. This implies that each additional unit of water saved for crop use will enhance production. For example, each hectare centimeter of water transpired by wheat in northern India added 200 kg of grain. The success of dryland crops and the efficiency of supplemental water in irrigated crops depend upon the water stored in the potential root zone. To achieve high soil water storage, it is essential to plug the pathways of water loss from soil.

Part of the water received at the soil surface as precipitation (P) and/or irrigation (I) infiltrates the soil, and part is lost as runoff (R). In equation form this process can be expressed as

$$P + I = R + IW$$

The relative proportions of infiltrated water (IW) and runoff are determined by the rate of water application and by soil characteristics. On a level field that is fully diked to hold the applied water, runoff is zero. But on sloping land, runoff occurs if the application rate exceeds infiltrability or if applied water exceeds water storage capacity of the soil or both. For conserving soil and increasing crop production, it is essential to increase the amount of infiltrated water. This is generally achieved by increasing infiltrability and/or by increasing the duration of contact between soil and water at the surface (termed *opportunity time*) by land shaping and other agronomic practices.

The infiltrated water is stored in the root zone and that exceeding the zone's storage capacity is lost as deep drainage (*D*) below the root zone. The stored water is lost as evaporation (*E*) from bare soil and as evapotranspiration (ET) from vegetated soil:

$$IW = S + D$$

and

$$S = \text{ET (or } E) + \Delta S$$

where *S* is water stored in the root zone, and ΔS is the change in stored water. Therefore, the total water balance of soil can be expressed as

$$P + I = R + D + \text{ET (or } E) + \Delta S \tag{1.1}$$

Under annual field crops, the soil surface remains bare during fallow, preparatory tillage, planting, germination, and seedling stages. Much water is lost during these periods by direct evaporation from soil. This is an unproductive loss and is often quite large. Wendt et al. (1970) reported that 40%–70% of precipitation in arid and semiarid regions was lost as evaporation. In a study of water balance for three cropping systems based on winter wheat (*Triticum aestivum* L.), evaporation during the fallow accounted for 36%–61% of precipitation (Stewart and Burnett, 1987). Even in growing crops, direct evaporation from the soil surface (E_s) constitutes a substantial part of ET. In row crops like corn (*Zea mays* L.), E_s was as high as 50% of ET during a normal season (Tanner, Peterson, and Love, 1960; Peters and Russel, 1959; Peters, 1960). Evaporation from soil not only reduces the amount of water stored in the soil but also renders soil saline by depositing at the surface the salts that are moved upward with the evaporating water. Therefore, it is of prime concern to reduce this loss in order to increase the storage of plant-available water in the root zone in bare fields and cause a greater fraction of ET to be used as transpiration in cropped fields.

The textbook information on evaporation from bare soil indicates that in the absence of a shallow water table, evaporation following wetting is a variable process. It takes place in three stages: *constant-rate stage, falling-rate stage,* and *low-rate stage.* In the constant-rate stage, the evaporation rate from soil equals that from a free water surface and is controlled mainly by atmospheric evaporativity (E_0). In the falling-rate stage, the evaporation rate is insensitive to E_0 and is dictated only by the water transmission properties of the soil. In the low-rate stage, a few centimeters of soil surface dry out to air-dry wetness and water is lost slowly as vapor movement through the dry layer. Also, the total evaporation under higher E_0 is always higher than that under lower E_0.

Research on ways to reduce soil evaporation dates back at least as far as the 1880s. Of the various means found useful for this purpose, the spreading or maintaining of crop residues at the soil surface as mulch appeared promising and economically feasible. Shallow tillage following wetting was reported to reduce evaporation from soil, but the benefit depended upon soil type, climatic factors, time of tillage after wetting, type of tillage, and the time of monitoring the benefit (Jalota and Prihar, 1990). The development of herbicides in recent decades has permitted an effective substitution of chemical for mechanical weed control. It has thus facilitated the retention of residues at the soil surface by eliminating the need to disturb the soil surface.

Although the application of crop residues at the soil surface is credited with reducing evaporation loss from soil (Jalota and Prihar, 1979; Greb, 1983; Unger, 1986; Brun et al., 1986), under certain situations it has been reported to enhance water loss from soil (McCalla and Army, 1961; Jacks, Brind, and Smith, 1955). There is a definite time pattern of evaporation reduction with residue mulching. Immediately after wetting, water loss from residue-covered soil is lower than that from bare soil, but as the drying proceeds the trend reverses and the benefit of residue cover starts declining. Since the extent and duration of surface wetness are influenced by rainfall, soil type, and amount and manner of placement of residue, these variables certainly modify the effect of crop residues on evaporation reduction.

Research in the last four decades has led to the questioning of existing doctrines on soil water evaporation and to the redefining of well-established concepts. Likewise, much more is now known about evaporation reduction by tillage and straw mulching. But this advanced knowledge is scattered in diverse journals, reports, and bulletins. Thus, in this book we have compiled the available information on the process of evaporation from soil and on direct and interactive impacts of tillage (time and type of tillage) and straw mulching (amount, type, and placement) on evaporation reduction as modified by soil type, E_0, and rainfall pattern.

In Chapter 2 we elaborate on the process of evaporation from free water and from bare, tilled, and residue-covered soils. In Chapter 3 we explain the procedures for measuring evaporation from the soil and the methods for estimating it using analytical, numerical, or purely empirical approaches. In Chapter 4 we discuss the benefits of tillage in reducing evaporation from the soil, as affected by soil type and evaporativity, and postdrying depth distribution of soil water in treated and untreated soil. In Chapter 5 we describe evaporation reduction by straw mulching in relation to soil type, evaporativity, and rainfall. This chapter also has a compilation of field observations on evaporation reduction and water use efficiency achieved by tillage and mulching during fallow and cropped periods. Finally, we deal with the effect of tillage and residue mulch combinations on evaporation reduction in Chapter 6.

References

Brun, L. J., J. W. Enz, J. K. Larsen, and C. Fanning. 1986. Springtime evaporation from bare and stubble-covered soil. *Journal of Soil and Water Conservation* 41:120–122.

Falkenmark, M., J. Lundqvist, and C. Widstrand. 1990. *Water Scarcity—An Ultimate Constraint in Third World Development.* Theme V Report 14, 1990. Linköping, Sweden: Department of Water and Environmental Studies, University of Linköping.

Greb, B. W. 1983. Water conservation: Central Great Plains. In *Dryland Agriculture,* edited by H. E. Dregne and W. O. Willis, pp. 57–72. Agronomy Monograph no. 23. Madison, Wis.: American Society of Agronomy.

Jacks, G. V., W. D. Brind, and R. Smith. 1955. *Mulching.* Bureau of Soil Technical Communication no. 49. Farnham Royal, Bucks, England: Commonwealth Agricultural Bureaux.

Jalota, S. K., and S. S. Prihar. 1979. Soil water storage and weed growth as affected by shallow tillage and straw mulching with and without herbicide in bare-fallow. *Indian Journal of Ecology* 5:41–48.

_____. 1990. Bare soil evaporation in relation to tillage. *Advances in Soil Science* 12:187–216.

McCalla, T. M., and T. J. Army. 1961. Stubble mulch farming. *Advances in Agronomy* 13:125–196.

Peters, D. B. 1960. Relative magnitude of evaporation and transpiration. *Agronomy Journal* 52:536–538.

Peters, D. B., and M. B. Russel. 1959. Relative water loss by evaporation and transpiration in field corn. *Soil Science Society of America Proceedings* 23:170–176.

Stewart, B. A., and E. Burnett. 1987. Water conservation technology and rainfed and dryland agriculture. In *Water Conservation and Water Policy in World Food Supplies,* edited by W. R. Jorden, pp. 355-359. College Station: Texas A & M University Press.

Tanner, C. B., A. L. Peterson, and J. R. Love. 1960. Radiant energy exchange in a corn field. *Agronomy Journal* 52:373–379.

Unger, P. W. 1986. Wheat residue management effects on soil water storage and corn production. *Soil Science Society of America Journal* 50:764–770.

Wendt, C. W., T. L. Olsen, H. J. Haus, and W. O. Willis. 1970. Soil water evaporation. In *Evapotranspiration in Great Plains.* Research Communication. Great Plains Agriculture Council Publication no. 50. Manhattan, Kans.: Kansas State University.

2

The Process of
Evaporation

Evaporation is the process of conversion of water from liquid to vapor phase and the escape of the latter into the atmosphere. It is an energy-controlled process. When (liquid) water molecules, which are in constant motion, attain sufficient momentum by absorbing latent heat of vaporization (2.4×10^6 J kg^{-1} at 20°C), the hydrogen bond is disrupted and the water molecules escape into the atmosphere. Thus, by virtue of its kinetic energy, a molecule changes from liquid phase to vapor phase. Simultaneously, molecules in vapor phase that lose energy may strike the (liquid) water surface and get absorbed. The relative rates of these two movements depend upon the concentration of vapor in the atmosphere relative to its concentration at the evaporating surface. Evaporation occurs when the number of molecules leaving the evaporating surface exceeds the number returning to it. The rate of evaporation is determined by (1) the difference between vapor pressures at the evaporating surface and in the turbulent atmosphere above and (2) the resistance to liquid and vapor flow in the system, which depends upon the nature of the flow medium. The resistance can be impedance to liquid and vapor flow in the system, the thickness of the laminar layer adjacent to the evaporating surface, or the resistance caused by the turbulent air (Figure 2.1). Thus, the evaporation function is analogous to Darcy's (1856) equation and can be expressed as

$E = -K' (de/dz)$

where E is evaporation rate,
z is the thickness of the laminar layer,

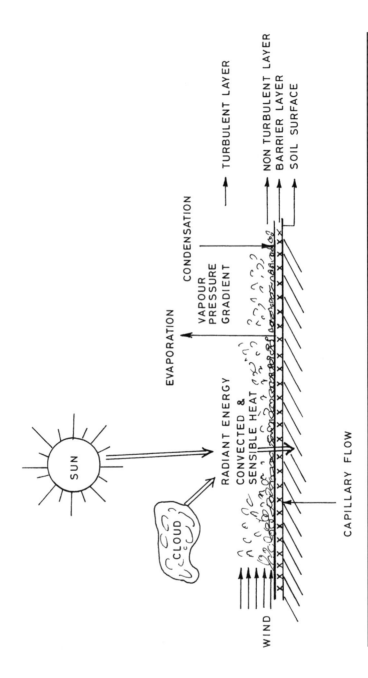

FIGURE 2.1. Schematic diagram of the process of evaporation from soil

de is the vapor pressure difference across dz, and
K' is the transfer coefficient.

Once vapor-phase water molecules enter the atmosphere, they are transported away by diffusion or convection or both, depending upon the air movement, which causes mixing of air layers.

Evaporation from a Free Water Surface

Dalton enunciated the fundamental law of evaporation from a free water surface in 1802 (see Ghildyal and Tripathi, 1987). This law states that evaporation will occur only if the actual vapor pressure of the air above the water surface is less than that at the water surface. The rate of evaporation was expressed as

$$E = (e_0 - e_d) \ f(U)$$

where E is evaporation rate,
e_0 is the mean vapor pressure at the water surface,
e_d is the mean vapor pressure in the air at some height above the water surface, and
$f(U)$ is a wind function whose value depends upon conductance of vapor from the surface to the measuring height.

Penman (1948) offered an empirical relationship for large surfaces. He found the conductance to be linearly related to wind speed at 2 m above the soil surface. Sweers (1976) reviewed many of the conductance formulae and concluded that conductance (h) could be best predicted as

$$h = 5.7 + 0.04u_3 \tag{2.1a}$$
$$h = 7.0 + 0.03u_{10} \tag{2.1b}$$

where u_3 and u_{10} in Equations 2.1a and 2.1b are the wind speeds (cm s^{-1}) at 3 m and 10 m above land surface, respectively. Under natural evaporative conditions, the conductance term needs some adjustments for atmospheric buoyancy (Tanner, 1968) to account for mass and momentum transfer across boundary layers. Obviously, accurate estimates of the conductance term are the major difficulty in describing the evaporation process (Sinclair, 1990). However, the function $f(U)$ is represented as $a + b(U)$ where a is the function of aerodynamic roughness, and b arises from the contribution of free thermal convection to the total transfer, which becomes important under calm conditions. Generally, b is small and is often taken as zero.

Factors Affecting Free Water Evaporation

Any meteorological factor that affects the supply of energy (latent heat of vaporization), the vapor pressure difference between the evaporating surface and the atmosphere, and the transport of vapor will influence the rate of evaporation. Radiation, temperature, relative humidity, and wind velocity are the meteorological factors influencing evaporation and together determine the evaporativity (E_0), which is defined as the maximum flux at which atmosphere can vaporize water from a free water surface (Hillel, 1971). Solar energy reaching the earth is the major factor influencing evaporation because it raises air temperature and provides the specific heat of water and the latent heat of vaporization. Temperature affects evaporation by increasing the vapor pressure gradient via increasing saturated vapor pressure at the water surface. An increase in atmospheric humidity decreases the vapor pressure gradient. Wind continually disturbs the air immediately above the soil surface and replaces air that may be fairly well saturated with water vapor with drier currents.

In nature, E_0 fluctuates diurnally (Equation 2.2) following a sine function (Hillel, 1976) and varies from day to day:

$$E_0 = E_{max} \; \sin \pi \; (t/86{,}400) \tag{2.2}$$

where E_{max} is midday evaporativity and t is time in seconds from sunrise. Evaporativity is always greater in summer than in any other season.

Evaporation from Untreated Bare Soil

Process and Time Trends

Evaporation from a soil surface is essentially the evaporation of water surrounding the soil particles as thin films and filling the pore spaces between them. Therefore, the atmospheric conditions that govern evaporation from a free water surface also govern evaporation from soil. Because the supply of water is limited in soil, compared with free water, and because the water molecules escaping from the soil have to overcome greater resistance than those escaping from a free water surface, due to their attraction by soil particles, evaporation from soil is also constrained by soil wetness. Thus, evaporation from soil has three basic physical requirements: (1) a source of energy to provide the latent heat of vaporization of water, (2) a vapor pressure gradient between the evaporating soil and the surrounding atmosphere, and (3) a supply of water to the evaporating surface (Hillel, 1971). The first two conditions— a supply of energy and removal of vapor—are external to the evaporating soil body and determine the potential rate of water loss (E_0). Removal of vapor increases with increasing temperature (T) and wind speed (WS) and decreases with increasing relative humidity (RH). In a 1994 study E_0 was found to be related to these variables through the following multiplicative power function (Equation 2.3), which explained 85% of the variability in E_0 in terms of these variables (Jalota, unpublished):

$$E_0 = 0.006 \text{WS}^{0.66} \ T^{\overset{**}{1.24}} \ \text{RH}^{-\overset{**}{0.37}} \quad (r^2 = .85) \tag{2.3}$$

The symbol ** denotes the significance at 1% probability. The E_0 values computed with Equation 2.3 agreed closely with those calculated with Penman's (1956) equation (Figure 2.2). In addition to E_0, evaporation from soil is also affected by the soil's thermal conductivity and albedo, which jointly affect the partitioning of energy available for evaporation. An increase in albedo decreases evaporation rate by reducing the energy load at the soil surface. In a simulation of a 10-day drying cycle for three albedo conditions, Hillel (1977) found albedo to be important for the first 3 days (Figure 2.3) of drying after wetting.

The third condition—a supply of water—is internal to the evaporating body. It depends upon the water retention and transmission properties of the

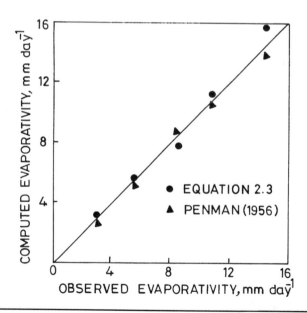

FIGURE 2.2. Comparison of observed evaporativities with those computed by Jalota's (unpublished) empirical relation and by Penman's (1956) equation

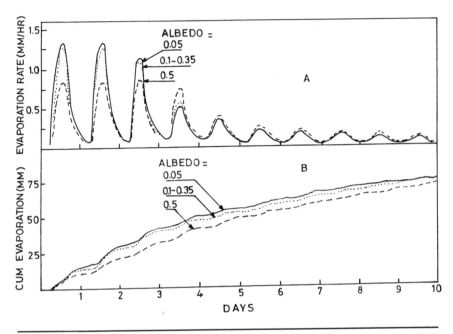

FIGURE 2.3.   Fluctuations in daily evaporation rate from a fine sandy loam soil with three values of soil reflectivity during 10 days of simulation (Hillel, 1977)

soil and is influenced by organic matter and salt content, the layering of the soil, and management practices such as tillage and compaction.

Evaporation from bare soil in the absence of a shallow water table is a variable process. Immediately following wetting by rain or irrigation, the soil has a high water content and can conduct water at high rates. The water content of the surface layer and the water-content-dependent transmission coefficient $K(\theta)$ decrease with time, but the potential gradient increases. Since water flux is the product of the hydraulic conductivity of soil and the hydraulic head gradient, which also change as a function of time, it is natural for evaporation rates to change with time. Early researchers (Stanhill, 1955; Lemon, 1956) divided the process of evaporation from soil into three stages. In the first stage, known as the *constant-rate stage*, the evaporation rate is governed by external conditions (air temperature, wind velocity, and relative humidity) and equals E_0 (Penman, 1941; Stanhill, 1955; Bond and Willis, 1970). Recent observations by Jalota (unpublished), however, have shown that the rate of evaporation from saturated soil may exceed that from a free water surface. For example, the ratio of evaporation from silt loam soil columns (95 cm long, 10 cm internal diameter) to that from similar-sized water columns exceeded

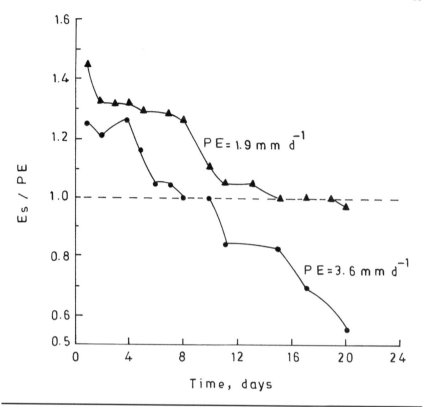

FIGURE 2.4. Ratio of maximum evaporation from saturated-soil (silt loam) columns to that from water-filled columns as a function of time under two evaporativities

unity for 15 and 8 days for constant free water under E_0 of 1.9 and 3.6 mm day^{-1}, respectively (Figure 2.4), in a drying chamber. E_0 in the chamber was achieved by using a thermostat and maintaining temperature at 29°C ± 0.5°C with incandescent light from electric bulbs equally spaced in the ceiling of the chamber and by controlling the humidity with a $CaCl_2$ solid:solution mixture at a number of places to ±3% and variation of wind speed with a table fan having a 40 cm sweep. Under these conditions the temperature of the soil columns was higher than that of the water columns (Figure 2.5). This phenomenon is attributed to the lower specific heat of the soil-and-water mix than that of water alone. The specific heat of dry soil is 0.2 cal g^{-1} °C^{-1} and that of water is 1.0 cal g^{-1} °C^{-1}. The greater temperature of the soil body increased the saturated vapor pressure (Figure 2.6) and consequently the vapor pressure gradient and the rate of evaporation. Earlier, Ghildyal and Tripathi (1987) ascribed the higher evaporation rate from saturated soil compared with that

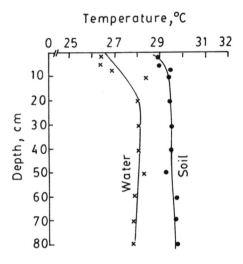

FIGURE 2.5. Temperature profiles of soil and water columns in the laboratory under an ambient temperature of 29°C ± 0.5°C

from free water to (innumerable) minute irregularities of the soil, which increased the area of the evaporating surface over that of free water.

After a certain threshold period, the rate of water loss from soil fell below E_0 and decreased with time. According to Gardner (1959) and Gardner and Hillel (1962), the evaporation rate during this *falling-rate stage* is controlled by the water transmission properties of the soil. They developed this concept from an analytical solution of the flow equation applied to uniformly wetted homogeneous soil, neglecting the effect of gravity and assuming instant drying of the soil surface to air dryness under infinitely high E_0:

$$\theta = \theta_0, \qquad x = 0, \qquad t > 0$$

where θ is the water content of soil,
θ_0 is the water content of air-dry soil,
x is soil depth, and
t is time.

But generally under field situations a surface soil layer a few centimeters thick dries out gradually under moderate E_0 and retains water in excess of air dryness for long periods. In such cases, E_0 could affect the evaporation rate (ER) during the early part of the falling-rate stage (Covey and Bloodworth, 1965). Jalota and Prihar (1986, 1987) showed both experimentally and theo-

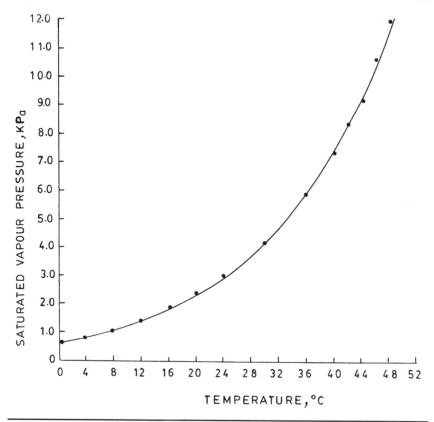

FIGURE 2.6. Saturated vapor pressure and temperature relations

retically that ER during the falling-rate stage was sensitive to E_0 (Figures 2.7-2.8). Data from an earlier report by Gardner and Gardner (1969) also support this contention.

More recently Jalota and Prihar (1991) reported that the duration of E_0 sensitivity of the ER during the falling-rate stage depended upon E_0 and soil type. They measured this duration by moving and matching the master curve based on the ER computed from the water transmission properties of soil using Equation 2.4, given by Gardner (1959), on the time (x) axis of the observed data:

$$ER = (\theta_i - \theta_0)\left(\frac{\overline{D}}{\pi t}\right)^{0.5} \tag{2.4}$$

The observed data points remained higher than those of the master curve until days 7 and 4 in the silt loam and days 5 and 2 in the sandy loam (Figure 2.9)

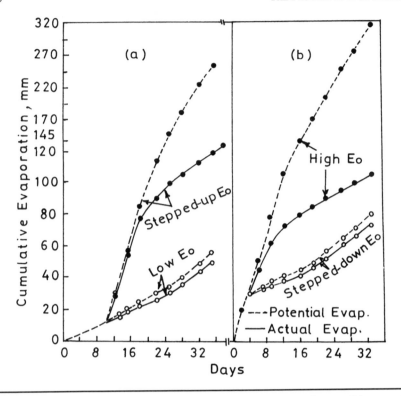

FIGURE 2.7. Cumulative evaporation as affected by (*a*) stepped-up or (*b*) stepped-down E_0 during the falling-rate stage from silt loam soil columns (Jalota and Prihar, 1986)

under evaporativities of 6.3 and 15.1 mm day⁻¹, respectively. The falling-rate stage had begun at 3 and 1 days under the respective evaporativities. In the loamy sand the observed ER matched the master curve from the beginning of the falling-rate stage (Figure 2.9). Apparently, in the silt loam, the influence of E_0 on ER continued for 4 days into the falling-rate stage under E_0 of 6.3 mm day⁻¹ and 3 days for 15.1 mm day⁻¹.

Jackson, Idso, and Reginato (1976) argued that during the initial period of the falling-rate stage (transition from energy-limited to soil-limited phase), the surface is not completely dried and the wet part evaporates at a rate dictated by both E_0 and the transmission properties of the soil. Mathematically the rate of evaporation is the product of soil water diffusivity and water content gradient. And since water content is strongly influenced by E_0 (Hillel, 1980), ER is bound to be influenced by E_0 during the early period of drying. Jalota and Prihar (1991) demonstrated that ER during the falling-rate stage became independent of E_0 only when the cumulative water loss from an initially wet (to

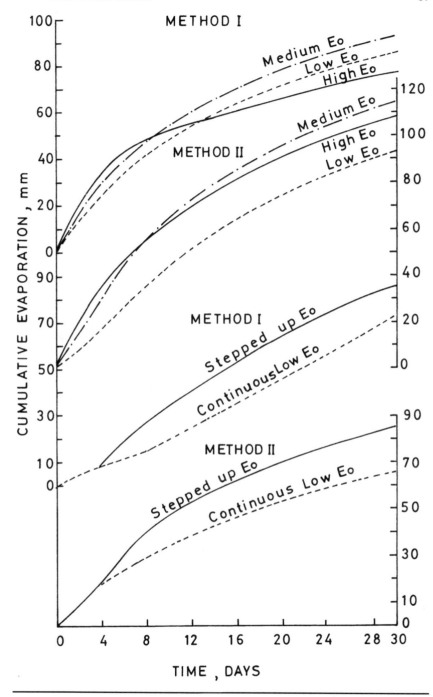

FIGURE 2.8. Time trends in predicted evaporation by two methods (see Chapter 3) under constant and stepped-up E_0 during the falling-rate stage in sandy loam soil (Jalota and Prihar, 1987)

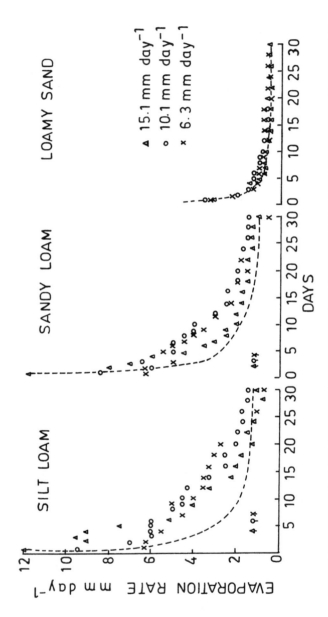

FIGURE 2.9. Evaporation rate as a function of time as affected by E_0 and soil type (Jalota and Prihar, 1991)

field capacity), semi-infinite (deep) profile, measured from the commencement of evaporation, reached a certain amount, irrespective of E_0. This amount was specific to soil type and was 40 mm for the silt loam and 23 mm for the sandy loam columns in their studies. Evidently, this amount of water loss increased with an increase in the water-retentive capacity of the soil. Gradually, a dry layer develops at the surface that acts as a barrier to vapor diffusion, and evaporation enters the third stage, called the *low-rate stage*. Here the water content in the surface layer reduces to air dryness, and ER is dependent upon the vapor diffusivity and vapor concentration in the dry surface-soil layer. Since the plane of evaporation shifts to below the surface, water transfer from there occurs only in the vapor state by molecular diffusion, and the flow is governed by the adsorptive forces acting over molecular distances at solid-liquid interfaces (Lemon, 1956).

The concept of three stages of drying does not hold under field conditions (Jackson, Idso, and Reginato, 1976; Minhas, 1985). Because of diurnal fluctuations in the evaporative demand, the surface layer absorbs moisture during the night, when temperature is low and soil water and soil temperature gradients are both in the downward direction and cause water to flow to the surface (Rose, 1968). Hillel (1977) reported that the water content of the surface layer increased by 1%-3% during the night. With sunrise, ER equals E_0 for some time, and then with the rise in temperature, the temperature gradient reverses and causes water vapor to move downward, with the result that the moisture content of the surface layer decreases and evaporation starts occurring as in the falling-rate stage. This directional opposition of vapor and liquid fluxes during daytime and their concurrence during nighttime determine the fluctuations in the diurnal evaporation rate from soil. Under such cyclic E_0, the alternating desorption and resorption of moisture by the soil surface inevitably involve soil moisture hysteresis, and the cumulative evaporation (CE) from soil tends to be lower than under equivalent steady E_0 (Hillel, 1976).

Total evaporation and evaporation rate during a given stage of evaporation are influenced by E_0, soil type, and redistribution time for water before commencement of evaporation (Jalota and Prihar, 1990). These variables, discussed later, affect evaporation trends by altering the energy supply to the site of evaporation, the rate of water supply within the soil, and the vapor pressure gradient between the evaporation site and the atmosphere.

Factors Affecting Evaporation from Untreated Bare Soil

Important factors affecting the course of evaporation from bare soil are evaporativity, soil characteristics, initial soil water content profile, salt content, and profile layering.

Evaporativity

The effect of evaporativity (E_0) on evaporation has been investigated by a number of workers. Penman (1941), Wiegand and Taylor (1960), Willis (1962), Gardner and Hillel (1962), Benoit and Kirkham (1963), Fritton, Kirkham, and Shaw (1967), and Acharya, Sandhu, and Abrol (1979) included more than one level of E_0 in their controlled experiments. Two divergent views emanated from these studies. Gardner and Hillel (1962) showed theoretically as well as experimentally that the cumulative evapora-tion (CE) with time always remained higher under higher E_0 than under lower E_0. The other point of view was that the initial higher evaporation rate under high E_0 may in the long run reduce total water loss to the atmosphere because of the rapid formation of the dry layer, which reduces later water loss (Covey and Bloodworth, 1965; Penman, 1941; Soviet work cited by Lemon, 1956). Strictly speaking, the CE from soil is dependent not only upon E_0 but also upon the evaporative conditions and the water retention and transmission properties of the soil.

The first view was based on the solution of the general flow equation. But the experimental results and numerical solutions of the flow equation (Jalota and Prihar, 1986) have shown that the effect of E_0 on CE depends upon the soil type and the initial wetness of the soil. When evaporation occurred from slow-draining soils, in which the water transmission coeffi-cient declined gradually with the decrease in water content that commenced immediately after wetting, the CE at any time up to 30 days remained higher under higher E_0 than under lower E_0 (Figure 2.10). However, in soils of low retentivity or where the soil surface water content was lowered by redistribution before commencement of evaporation, CE under high E_0 fell below that under lower E_0 after a certain period. Further, the evaporation rate remained higher under higher E_0 than under lower E_0 as long as the decrease in soil water diffusivity under higher E_0 was accompanied by a large increase in soil water content gradient. When the product of diffusiv-ity and water content gradient under higher E_0 fell below that under lower E_0, the evaporation rates under higher and lower E_0 reversed. When this con-tinued for a sufficiently long time, CE under higher E_0 lagged behind that under lower E_0 (Jalota and Prihar, 1987). Hanks, Gardner, and Fairbourn (1967) reported that for comparable evaporativities, drying by wind caused higher CE than drying by radiation. Under radiation conditions the warm-ing of the soil surface raises the vapor pressure and causes thermally induced downward flow, which counters the upward water flow due to the moisture gradient. The effect of E_0 on evaporation rates during the falling-rate stage has been discussed earlier.

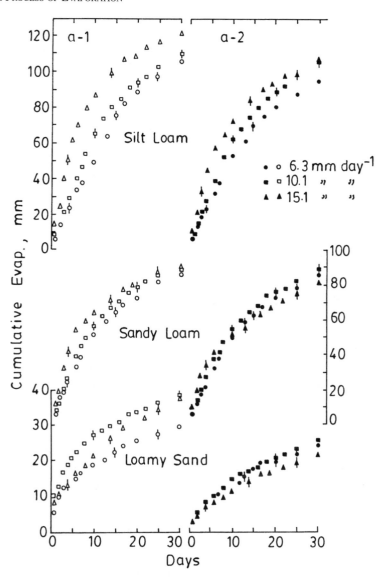

FIGURE 2.10. Cumulative evaporation from silt loam, sandy loam, and loamy sand immediately after wetting (*hollow symbols*) and with 2-day redistribution time (*solid symbols*) for low, medium, and high evaporativities (Jalota and Prihar, 1986)

Soil Type

Under a given E_0 the soil type affects CE through its retention and trans-mission properties. In the wet range the transmission rate is higher in coarse-textured than in fine-textured soils, while the reverse is true in the dry range (Wind, 1961). Therefore, at the beginning of evaporation from a nearly satu-rated soil, which holds moisture at very low tensions, evaporation is initially higher from coarse soil (Penman, 1941; Prihar, Singh, and Sandhu, 1968); the trend reverses after some time (Gill et al., 1977). On the other hand, when evap-oration starts from a soil with initial moisture content at or near field capacity, the evaporation rate in the initial as well as later stages remains higher in fine-textured soils than in coarse-textured soils (Jalota and Prihar, 1986).

Redistribution Time

Redistribution of water prior to commencement of evaporation or occurring simultaneously with evaporation reduces evaporation losses. This is so because it decreases the water content and hence the overall gradient and soil water dif-fusivity in the upper zone subject to evaporation. Reduction in CE with post-irrigation redistribution has been reported under both laboratory (Staple, 1976) and field (Jaggi, Gupta, and Russel, 1978) conditions. In 45 cm deep soil columns dried under high E_0, the greater redistribution of water added slowly to the soil columns resulted in lower CE than when the same amount of water was added by ponding over the surface, which infiltrated rapidly (Bresler, Kemper, and Hanks, 1969). More recently Jalota and Prihar (1986) showed that the effect of redistribution time on evaporation reduction was influenced by soil texture—the $K(\theta)$ relation—and the prevailing E_0. Redistribution brought about a greater reduction in evaporation in coarse-textured soils than in fine-textured soils (Figure 2.10), which is attributed to faster drainage and lower retention of water in the surface layers of the coarse soil. In coarse soils water moved into lower layers and became less susceptible to evaporation. Lower water contents in surface layers result in earlier formation of the dry layer, especially under high E_0, which further reduces the evaporation losses.

Evaporation from Tilled Soil

The loosening of surface soil by tillage causes significant changes in its energy reflection and water transmission properties, and therefore, the course of evaporation from tilled soil differs from that of untilled soil. Linden (1982) pointed out that loosening the soil increases surface roughness, which reduces albedo and increases potential evaporation by concentrating heat in the surface layers (Allmaras, Nelson, and Hallauer, 1972; Bowers and Hanks, 1965; Cary and Evans, 1975; Idso, Jackson, and Reginato, 1975; van Bavel and Hillel, 1976). Roughness also increases the surface area of soil exposed to the atmosphere and allows greater wind penetration (Ojeniyi and Dexter, 1979), both of which cause increased evaporation. The hydraulic properties of soil are also altered by tillage. Bulk density and pore size changes associated with loosening cause soil suction to change at a given volume water fraction and, thus, change hydraulic conductivity (Box and Taylor, 1962; Archer and Smith, 1972). A decrease in bulk density increased retention at low suction and decreased it at high suction (Laliberte and Brooks, 1967). Similarly, a decrease in bulk density increased hydraulic conductivity at low suction and decreased it at high suction.

Decreased albedo, increased surface area exposed to the atmosphere, and greater penetration by the wind into the loose and rough surface layer of tilled soil all increase the initial rate of evaporation. Water is rapidly lost from the upper portion of this layer and is replenished more slowly in the tilled than in the untilled soil because of the decreased hydraulic conductivity of the loosened layer. Hence, the top few centimeters of the tilled layer dry out more rapidly and thoroughly than the corresponding layer of the untilled soil. This dry layer acts as a barrier to both liquid and vapor flow to the surface and reduces further evaporation. This decrease is greater in tilled than in untilled soil because the untilled soil permits liquid flow from the interior of the soil to the surface for a longer period. The breaking of capillaries by tillage retards the hydraulic conductivity or diffusivity of the profile, particularly of the surface zone. However, the liquid flow across the tilled layer is not completely cut off. In fact, the lower portion of the tilled layer may for some time permit upward capillary flow to some distance above the interface, depending upon the fineness of clods, the water content of the tilled layer, and the type of soil. Gill and Prihar (1988) presented evidence of liquid water flow from untilled soil to seeds placed some distance above in the (simulated) tilled layer. When the tilled layer was separated from the untilled soil below by a thin layer of waterproofed aggregates, which permitted only vapor flow, water flow to the seeds was reduced. This implied that in the absence of a waterproofed layer, the liquid water moved to the seeds through the loose soil. Jalota and Prihar (1992) verified this phenomenon and reported that liquid flux across the interface of tilled and untilled layers in 30 days of drying (after tillage) constituted 72% and 42% of total water loss under E_0 of 3.6 mm day^{-1} and 15.6 mm day^{-1}, respectively, in silt loam soil and 20% and 30% in sandy loam soil. Evidently, therefore, the phase change from liquid water to vapor in the tilled soil does not occur at the interface of tilled and untilled soil but at various loci within the tilled layer. Nevertheless, formation of the dry layer (1) moves the locus of conversion of liquid water to vapor some (variable) distance below the surface and (2) lowers the temperature at the interface of tilled and untilled layers (Benoit and Kirkham, 1963) and, thus, reduces the saturated vapor pressure at that plane. The dry layer also resists the flow of vapor from soil to atmosphere and conducts liquid water much slower than the undisturbed soil below.

With time, the surface layer of the untilled soil also dries out and the locus of phase change moves deeper. The dry layer that develops at the surface of the undisturbed soil is less porous than that created by tillage. When the drying front moves deeper into the untilled soil, the less porous dry surface layer permits less vapor flow compared with an equal depth of the more porous tilled layer (Acharya and Prihar, 1969). Consequently, after a certain time, the rate of evaporation loss from the tilled soil begins to exceed that from the untilled. The time course of evaporation flux from tilled soil normalized against that from untilled soil is shown schematically in Figure 2.11. For a short time after tillage, the evaporation rate from tilled soil exceeds that from untilled. With the rapid formation of a dry layer by the accelerated drying of the tilled surface, evaporation from tilled soil lags behind that from the untilled. Eventually, when a dry layer also forms at the surface of untilled soil, the evaporation rate from tilled soil again exceeds that from untilled.

The duration and magnitude of this fluctuation in evaporation rates in tilled versus untilled soil, which determines the course of evaporation reduction with tillage, have been found to depend on soil type, E_0, time and type of tillage, and their interactions (see Chapter 4).

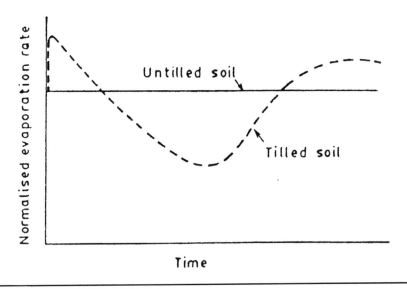

FIGURE 2.11.   Schematic representation of evaporation from tilled soil relative to that from untilled soil (Jalota and Prihar, 1990)

Evaporation from Crop-Residue-Covered Soil

Application of crop residues at the soil surface alters the magnitude of the prerequisites for evaporation and causes the evaporation trend of residue-covered soil to deviate from that of a bare soil. The evaporation rate from residue-covered soil normalized against that from bare soil is shown schematically as a function of time in Figure 2.12. Initially, the evaporation rate from residue-covered soil is lower than that from bare soil. This is so because the residue reduces (1) energy reaching the soil surface and vapor pressure at the soil surface by reflecting or intercepting the solar radiation and (2) vapor diffusion from the site of evaporation to the atmosphere by increasing the thickness of nonturbulent air above the soil surface (Army, Wiese, and Hanks, 1961).

The duration of this lower evaporation rate depends upon soil type, E_0, wetting pattern, and residue characteristics (type, amount, and manner of application) and their interactions. The period of lower evaporation rate is longer for fine-textured soils under lower E_0 that have plentiful residue cover and are subject to more frequent rains. As the residue cover lowers the evaporation rate, it

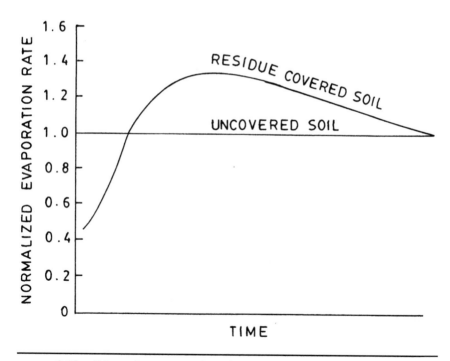

FIGURE 2.12.   Schematic representation of evaporation from residue-covered soil relative to that from bare soil

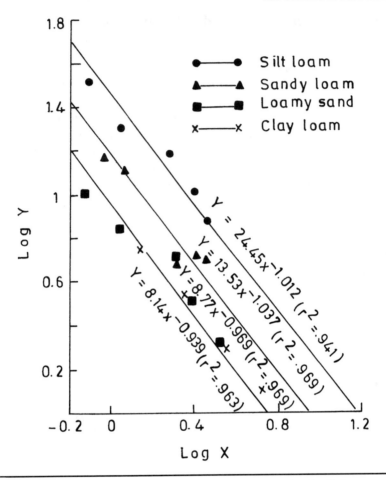

FIGURE 2.13.   Duration of the constant-rate stage of evaporation in silt loam, sandy loam, and loamy sand soils (Prihar, Jalota, and Steiner, 1996)

increases the duration of the constant-rate stage in accordance with the following relation obtained by Bond and Willis (1970) for a fine sandy loam soil:

$$\log Y = 1.779 - 1.233 \log X \quad (r^2 = .99) \tag{2.5}$$

where Y is the duration of the constant-rate stage in days, and X is the evaporation rate (mm day^{-1}). Recent studies by Prihar, Jalota, and Steiner (1996) have shown that this relation is soil specific and that the duration of the constant-rate stage for a given evaporation rate increases with the fineness of soil texture (Figure 2.13). A delayed transition from the first to the second stage of evaporation with residue was also reported by Massee and Cary

(1978). The higher initial evaporation rate from bare soil compared with that from residue-covered soil lowers the water content of the surface layer, with a concomitant decrease in unsaturated hydraulic conductivity and hence evaporation rate. As evaporation proceeds, a time comes when the evaporation rate from bare soil equals that from residue-covered soil. At this point the magnitude of the forces resisting evaporation in the residue-covered soil permits an evaporation rate equal to that of the product of decreased diffusivity and increased water content gradient in the bare soil as a result of surface drying.

As the surface drying of bare soil continues, a dry layer in the form of a natural soil mulch is formed at the surface. The evaporation rate from the bare soil falls below that from the residue-covered soil because the surface soil remains wet under the residue and water is supplied from lower layers to the surface for longer periods (Minhas and Gill, 1985). Thus, the dynamics of the relative evaporation rates from residue-covered and bare soil determines the time course of evaporation reduction with residue mulches.

Summary

Evaporation from soil in the absence of a shallow water table is a variable process that early researchers divided into three distinct stages: constant-rate stage, falling-rate stage, and low-rate stage. However, a critical analysis of the literature discussed in this chapter shows that the situation is more complex. The transition from one stage to the other depends upon soil type, atmospheric evaporativity (E_0), redistribution of the applied water before commencement of evaporation, time and type of tillage, and the amount, type, and position of residue at the soil surface. For example, in coarse-textured soils the constant-rate stage is practically absent, especially when E_0 is high. Contrary to this, in finer-textured soils under low E_0 when the surface is covered with crop residue, this stage lasts for a long period and the falling-rate stage is entered more gradually.

Recent studies have shown that we need to refine some earlier concepts. The assumption that evaporation from soil is either equal to or less than that from a free water surface has been disproved for fine-textured soils and for turbulent wind conditions. This is so because the specific heat of soil is lower than that of free water, and its surface area is larger. Similarly, contrary to the earlier doctrine, the evaporation rate during the falling-rate stage has been found to be sensitive to E_0. Previously it was held that the evaporation rate during the falling-rate stage was governed only by the water transmission properties of soil and was not affected by E_0. Interestingly, the evaporation rate from soil becomes independent of E_0 when the soil has lost a specified amount of water regardless of E_0. This amount is greater in fine-textured than in medium-textured soils. Also, the previous contention that cumulative evaporation from soil is always higher under higher E_0 than under lower E_0 does not hold for all situations.

The evaporation rate is initially greater in tilled than in untilled soils because the increased surface area and roughness of the tilled surface layer increase the energy supply at the soil surface in terms of wind gustiness and temperature. Once the tilled layer dries out, the evaporation rate is reduced, with the size of the reduction dependent upon the layer's porosity and thickness. In residue-covered soil the evaporation rate remains less than that from bare soil for a variable length of time depending upon the water transmission properties of the soil, E_0, and amount, type, and position of cover. Under prolonged drying when a dry layer develops at the surface of bare soil, evaporation from residue-covered soil may exceed that from bare soil. These observations warrant that practices to reduce soil water evaporation (tillage or straw mulching) should be carefully chosen in relation to soil type, evaporativity, frequency and amount of rain, and the nature and amount of available residue and its manner of application.

References

Acharya, C. L., and S. S. Prihar. 1969. Vapor losses through soil mulch at different wind velocities. *Agronomy Journal* 61:666–668.

Acharya, C. L., S. S. Sandhu, and I. P. Abrol. 1979. Effect of exchangeable sodium on the field drying pattern of soil at two evaporativities. *Soil Science* 127:56–62.

Allmaras, R. R., W. W. Nelson, and E. A. Hallauer. 1972. *Fall vs. Spring Plowing and Related Heat Balance in the Western Corn Belt*. Minnesota Agricultural Experiment Station Technical Bulletin 283. Saint Paul.

Archer, J. R., and P. D. Smith. 1972. The relation between bulk density, available water capacity, and air capacity of soils. *Journal of Soil Science* 23:475–480.

Army, T. J., A. F. Wiese, and R. J. Hanks. 1961. Effect of tillage and chemical weed control practices on soil moisture losses during fallow period. *Soil Science Society of America Proceedings* 25:410–413.

Benoit, G. R., and D. Kirkham. 1963. The effect of soil surface conditions on evaporation of soil water. *Soil Science Society of America Proceedings* 27:495–498.

Bond, J. J., and W. O. Willis. 1970. Soil water evaporation: First stage drying as influenced by surface residue and evaporation potential. *Soil Science Society of America Proceedings* 34:924–928.

Bowers, S. A., and R. J. Hanks. 1965. Reflection of radiant energy from soils. *Soil Science* 100:136–138.

Box, J. E., and S. A. Taylor. 1962. Influence of soil bulk density on matric potential. *Soil Science Society of America Proceedings* 26:119–123.

Bresler, E., W. D. Kemper, and R. J. Hanks. 1969. Infiltration, redistribution, and subsequent evaporation of water from soil as affected by wetting rate and hysteresis. *Soil Science Society of America Proceedings* 33:832–840.

Cary, J. W., and D. D. Evans. 1975. *Soil Crusts*. Arizona Agricultural Experiment Station Technical Bulletin 214. Tucson.

Covey, W., and M. Bloodworth. 1965. Evaporativity and second stage of drying of soil. *Journal of Applied Meteorology* 5:364–365.

Darcy, H. 1856. *Les fontaines publique de la ville de Dijon*. Paris: Dalmont.

Fritton, D. D., D. Kirkham, and R. H. Shaw. 1967. Soil water and chloride redistribution under various evaporation potentials. *Soil Science Society of America Proceedings* 31:599–603.

Gardner, H. R., and W. R. Gardner. 1969. Relation of water application to evaporation and storage of water. *Soil Science Society of America Proceedings* 33:192–196.

Gardner, W. R. 1959. Solutions of flow equation for the drying of soils and other porous media. *Soil Science Society of America Proceedings* 23:379–382.

Gardner, W. R., and D. I. Hillel. 1962. The relation of external evaporation conditions to the drying of soils. *Journal of Geophysical Research* 67:4319–4325.

Ghildyal, B. P., and R. P. Tripathi. 1987. Evaporation. In *Soil Physics*, pp 401–434. New Delhi: Wiley Eastern.

Gill, K. S., S. K. Jalota, S. S. Prihar, and T. N. Chaudhary. 1977. Water conservation by soil mulch in relation to soil type, time of tillage, tilth, and evaporativity. *Journal of the Indian Society of Soil Science* 25:360–366.

Gill, K. S., and S. S. Prihar. 1988. Seedling emergence at four temperatures from drying out seed zones underlain by wet soil. *Plant and Soil* 112:267–272.

Hanks, R. J., H. R. Gardner, and M. L. Fairbourn. 1967. Evaporation of water from soil as influenced by drying with wind or radiation. *Soil Science Society of America Proceedings* 31:593–598.

Hillel, D. 1971. *Soil and Water Physical Principles and Processes*, pp. 184, 192–200. New York: Academic Press.

———. 1976. On the role of soil moisture hysteresis in the suppression of evaporation from bare soil under diurnally cyclic evaporativity. *Soil Science* 122:309–314.

———. 1977. *Computer Simulation of Soil-Water Dynamics: A Compendium of Recent Work*. International Development Research Center, 60 Queen Street, Ottawa, Canada.

———. 1980. *Fundamentals of Soil Physics*. New York: Academic Press.

Idso, S. B., R. D. Jackson, and R. J. Reginato. 1975. Estimating evaporation: A technique adaptable to remote sensing. *Science* 189:991–992.

Jackson, R. D., S. B. Idso, and R. J. Reginato. 1976. Calculation of evaporation rate during the transition from energy-limiting to soil-limiting phases using albedo data. *Water Resources Research* 12:23–26.

Jaggi, I. K., R. K. Gupta, and M. B. Russel. 1978. Evaporation from bare soil. *Journal of the Indian Society of Soil Science* 26:71–73.

Jalota, S. K., and S. S. Prihar. 1986. Effect of atmospheric evaporativity, soil type, and redistribution time on evaporation from bare soil. *Australian Journal of Soil Research* 24:357–366.

_____. 1987. Observed and predicted evaporation trends from a sandy loam soil under constant and staggered evaporativity. *Australian Journal of Soil Research* 25:243–249.

_____. 1990. Bare soil evaporation in relation to tillage. *Advances in Soil Science* 12:187–216.

_____. 1991. Evaporativity-sensitive evaporation during falling-rate stage as influenced by soil texture. *Journal of the Indian Society of Soil Science* 39:409–414.

_____. 1992. Liquid component of evaporative flow in two tilled soils. *Soil Science Society of America Journal* 56:1893–1898.

Laliberte, G. E., and R. H. Brooks. 1967. Hydraulic properties of disturbed soil material affected by porosity. *Soil Science Society of America Proceedings* 31:447–454.

Lemon, E. R. 1956. Potentialities for decreasing soil water loss. *Soil Science Society of America Proceedings* 20:160–163.

Linden, D. R. 1982. Predicting tillage effects on evaporation. In *Predicting Tillage Effects on Soil Physical Properties and Processes*, pp. 117–132. ASA Special Publication no. 44. Madison, Wis.: American Society of Agronomy, Soil Science Society of America.

Massee, T. W., and J. W. Cary. 1978. Potential for reducing evaporation during summer fallow. *Journal of Soil and Water Conservation* 33:126–129.

Minhas, P. S. 1985. Salt and water movement in unsaturated soils as affected by method of water application, surface soil condition, and atmospheric evaporativity. Ph.D. diss., Punjab Agricultural University, Ludhiana, India.

Minhas, P. S., and A. S. Gill. 1985. Evaporation from soil as affected by incorporation and surface and sub-surface placement of oat straw. *Journal of the Indian Society of Soil Science* 33:774–778.

Ojeniyi, S. O., and A. R. Dexter. 1979. Effect of soil structure and meteorological factors on temperature in tilled soils. In *Soil Physical Properties and Crop Production in the Tropics*, edited by R. Lal and D. J. Greenland, pp. 273–283. New York: Wiley.

Penman, H. L. 1941. Laboratory experiments on evaporation from fallow soil. *Journal of Agricultural Science* 31:454–463.

_____. 1948. Natural evaporation from open water, bare soil, and grass. *Proceedings of the Royal Society*, London 93:120–195.

_____. 1956. An introductory survey. *Netherlands Journal of Agricultural Sciences* 4:9–29.

Prihar, S. S., S. K. Jalota, and J. L. Steiner. 1996. Residue management for evaporation reduction in relation to soil type and evaporativity. *Soil Use and Management* 12:150–157.

Prihar, S. S., B. Singh, and B. S. Sandhu. 1968. Influence of soil and environment on evaporation losses from mulched or unmulched pots. *Journal of Research, Punjab Agricultural University* (Ludhiana, India), 5:320–328.

Rose, C. W. 1968. Evaporation from bare soil under high radiation conditions. In *Transactions of the 9th International Congress of Soil Science*, pp. 57–68. Adelaide, Australia.

Sinclair, T. H. 1990. Theoretical considerations in the description of evaporation and transpiration. In *Irrigation of Agricultural Crops*, edited by B. A. Stewart and D. R. Nielsen, pp. 339-360. Agronomy Monograph no. 30. Madison, Wis.: American Society of Agronomy.

Stanhill, G. 1955. Evaporation of water from field conditions. *Nature* (London) 176:82–83.

Staple, W. J. 1976. Prediction of evaporation from columns of soil during alternate periods of wetting and drying. *Soil Science Society of America Journal* 40:756–761.

Sweers, H. E. 1976. A nomograph to estimate the heat exchange coefficient at the air-water interface as a function of wind speed and temperature: A critical survey of some literature. *Journal of Hydrology* 30:375–401.

Tanner, C. B. 1968. Evaporation from plants and soil. In *Water Deficits and Plant Growth*, edited by T. T. Kozlowski, vol. 1, pp. 73–106. New York: Academic Press.

van Bavel, C. H. M., and D. I. Hillel. 1976. Calculating potential and actual evaporation from a bare soil surface by simulation of concurrent flow of water and heat. *Agricultural Meteorology* 17:453–476.

Wiegand, C. A., and S. A. Taylor. 1960. The temperature dependence of the drying of soil columns. In *Transactions of the 7th International Congress of Soil Science* (Madison, Wis.), vol. 1, pp. 169–177. Wageningen, the Netherlands: International Society of Soil Science.

Willis, W. O. 1962. Effect of partial covers on evaporation from soil. *Soil Science Society of America Proceedings* 26:598–601.

Wind, G. P. 1961. *Capillary Rise and Some Applications of the Theory of Moisture in Unsaturated Soil*. Institute of Land and Water Management Research Technical Bulletin 22. Wageningen, the Netherlands.

3

Measurement and Modeling of Soil Evaporation

Quantitative estimates of potential and actual evaporation losses from soil are often required by agronomists, hydrologists, meteorologists, and others for various purposes. For example, such estimates are essential for assessing soil water storage and distribution in the profile, relating evaporation loss with parameters determining the rates of evaporation, and assessing the effect of treatments intended to alter the magnitude of these losses. Direct measurement of evaporation losses is difficult with existing methods. Evaporation losses are, therefore, measured indirectly. Evaporation loss is one of the components of total water balance (see Equation 1.1). If all components of the balance (amount of rainfall and irrigation, on the one hand, and runoff, deep drainage, and change in water storage, on the other) are measured, the evaporation loss can be computed as the difference between the two sides. Under field conditions, evaporation is normally determined in this manner. But the existing methods for determining runoff, deep drainage, spatial variability of soil properties, and distribution of rain and water under field conditions do not permit precise estimates of these components. Consequently, scientists have resorted to simpler alternatives for estimating evaporation from soil. They use lysimeters, which are blocks of soil that are separated hydrologically from the surrounding soil but yet behave as part of the soil system. Lysimeters vary in size but permit periodic or continuous weighment. Because most lysimeters do not permit runoff and deep drainage, evaporation is equated with the loss in mass after allowing evaporation to occur. When deep drainage is allowed from the lysimeter, evaporation predictions based on the mass loss from the lysimeter are more accurate.

Prihar and van Doren (1967) used "baby lysimeters" that were 0.15 m in diameter and 0.25 m deep to determine the effect of shallow cultivation on direct evaporation from soil. The lysimeters were embedded between corn rows in tar-paper-lined holes. Boast and Robertson (1982) evaluated much smaller "microlysimeters" (a few centimeters in diameter and depth), which could be removed temporarily from their container holes for weighment to compute evaporation loss. Boast and Robertson mentioned that such devices were used as far back as the 1870s. But measurements from such small lysimeters do not represent real conditions. Boast and Robertson (1982) studied the systematic deviation of microlysimeters from reality to calculate the relation between lysimeter depth and the length of time they could be used to measure evaporation from soil.

Larger lysimeters of weighing type or floating type have also been used to determine evaporation or evapotranspiration losses (Pruit and Angus, 1960; McIlroy and Angus, 1964; van Bavel and Myers, 1962). According to Boast and Robertson (1982): "the validity of lysimetric methods for determining evaporation hinges on whether the evaporation from the isolated body of soil is essentially the same as that from a comparable non-isolated body. A number of factors can cause a deviation from reality: changes in the hydrological boundaries, imposition of a plane of zero flow, disturbance of soil during construction, conduction of heat by lateral walls etc." Large lysimeters are expensive to build and also difficult to move. Undisturbed or reconstituted soil columns have also been used for measuring evaporation under controlled laboratory conditions. Despite their shortcomings, lysimeters and soil column evaporation measurements are taken as standard for comparison with the modeled evaporation.

Modeling is another tool for evaporation assessment. It ranges from purely empirical regression equations to mathematical solutions of equations concerning the soil-water-atmosphere system. The mathematical solutions involve two approaches: analytical and numerical. The solutions using these approaches for varying soil and atmospheric conditions are the work of different researchers and are described here.

Evaporation from Untreated Bare Soil

Analytical Approach

Time Trends in Evaporation Rates

The model describing one-dimensional soil water flow in unsaturated homogeneous soil consists of a set of nonlinear equations. The principal equation derived from the theory based on total potential ($\phi = h + z + \Omega$),

Darcy's law ($v = -k\nabla_z \phi$), and the continuity equation ($\delta\theta/\delta t = -\nabla_z v$) is known as Richard's equation:

$$\frac{\delta\theta}{\delta t} = \frac{\delta}{\delta z}\left[D(\theta)\frac{\delta\theta}{\delta z}\right] \qquad (3.1)$$

where ϕ is hydraulic potential (mm),
h is matric potential (mm),
Ω is overburden potential (mm),
k is hydraulic conductivity (mm s^{-1}),
v is water flux (mm^3 mm^{-2} s^{-1}),
∇ is the del operator, taken here only vertically since flow is considered to be one-dimensional,
θ is soil water content (mm^3 mm^{-3}),
t is time (s),
z designates the vertical coordinate (positive downward) (mm), and
$D(\theta)$ is the hydraulic diffusivity (mm^2 s^{-1}).
Note that the gravity effect is neglected.

For constant-rate stage evaporation Gardner and Hillel (1962) solved the flow equation (Equation 3.1) analytically, subject to the following boundary conditions:

$\theta = \theta_i$ for $0 < z < L$, $t = 0$

$\dfrac{\delta\theta}{\delta z} = 0$ for $z = L$, $t > 0$

$D\dfrac{\delta\theta}{\delta z} = E$ for $0 < z < L$, $t > 0$

$\theta = \theta_c$ for $0 < z < L$, $t = t_c$

$\theta = \theta_a$ for $z = 0$, $t = t_c$

The duration of the constant-rate stage (t_c, in days) and the soil water content of the profile at the end of the stage (θ_c) are obtained as follows:

$$t_c = (\theta_i - \bar{\theta}_c)\frac{L}{\beta}$$

$$\bar{\theta}_c = L\left[\theta + \frac{1}{\beta}\ln\left(1 + \frac{E\beta L}{2D_a}\right)\right]$$

where θ_i is initial soil water content (mm^3 mm^{-3}),

E is the evaporation rate (mm day^{-1}),

L is the length of the column (mm),

θ_a is air-dryness water content (mm^3 mm^{-3}),

D_a is soil water diffusivity at θ_a (mm^2 day^{-1}), and

β is the constant of the function $D(\theta) = D_a \exp[\beta(\theta - \theta_a)]$.

Covey (1963) made a mathematical study of the constant-rate stage drying for finite and semi-infinite homogeneous soil columns and developed an index, G, that depended upon the soil type and atmospheric evaporativity as follows:

$$G = \beta q_0 \frac{L}{D_i}$$

where β is a constant for a given soil (cm^3 cm^{-3}),

D_i is diffusivity at initial soil water content (cm^2 s^{-1}),

L is the length of the column (cm), and

q_0 is evaporative flux (cm^3 cm^{-2} s^{-1}).

He reported that the value of G does not reflect the drying pattern in the initial stages, because all columns dry as semi-infinite. For longer drying, however, smaller values of G indicated slower drying of soil columns nearly uniformly with depth, and larger values of G meant more rapid drying of the exposed surface.

Gardner (1959) solved Equation 3.1 for isothermal drying of semi-infinite homogeneous soil columns during the falling-rate stage subject to the following assumptions: (1) the initial soil water content was uniform with depth at the start of drying, (2) evaporativity was infinite, (3) soil water diffusivity, $D(\theta)$, changed exponentially with soil water content (Gardner and Mayhugh, 1958), (4) the water content at the surface decreased to air dryness instantaneously, and (5) water content remained constant at the lower boundary of a semi-finite slab. Thus, the initial and boundary conditions were

$\theta = \theta_i$ for $z > 0$, $t = 0$

$\theta = \theta_a$ for $z = 0$, $t > 0$

He obtained Equation 3.2 for cumulative evaporation (CE) during the falling-rate stage:

$$CE = 2(\theta_i - \theta_a)\left(\frac{\overline{D}t}{\pi}\right)^{0.5} \tag{3.2}$$

where θ_i is initial soil water content,
θ_a is the air-dryness value of soil water content,
t is time,
π is a constant, and
$\overline{D}$ is mean weighted diffusivity, computed as

$$\overline{D} = \frac{1.85}{(\theta_i - \theta_a)^{0.85}} \int_{\theta_a}^{\theta_i} D(\theta)(\theta - \theta_a)^{0.85} d\theta$$

Gardner and Hillel (1962) assumed constant diffusivity with depth in addition to infinite E_0, uniform wetness, and neglect of gravity and obtained the following solution for the falling-rate stage in a finite system:

$$E = \frac{D(\overline{\theta})W\pi^2}{4L^2}$$

where E is the evaporation rate and $\overline{\theta}$ is the average soil water content obtained by dividing the total amount of water stored in the wetted zone (W) by the depth of wetting (L). The assumption of constant diffusivity with depth seems unrealistic. Similarly, the assumptions concerning infinite E_0, uniform wetness, and neglect of gravity in the solutions provided by Gardner (1959) and Gardner and Hillel (1962) also seem far from real. The gravity term can be neglected for fine-textured soils but not for coarse-textured ones, because with gravity water gets redistributed more rapidly into deeper layers and becomes less accessible for evaporation from the soil surface. The effect of gravity is of greater consequence when E_0 is high and the soil texture is coarse because it significantly decreases the evaporation loss (Jalota and Prihar, 1986). Fritton, Kirkham, and Shaw (1967) reported that the isothermal flow equation describes cumulative evaporation for both radiation and wind-drying treatments but cannot describe the water distribution pattern in the case of nonisothermal flow. The isothermal equation does not predict the formation of a dry surface layer. Such a layer could be predicted only when separate heat and mass transfer equations were used, as done by Rose (1968) for predicting evaporation from bare soil under high radiation. He established a theory that permits the calculation of evaporation from bare soil from water content and temperature profiles for the conditions of fairly high radiation in which diurnal fluxes interact with and modify the moisture flux. He gave

$$\langle E \rangle (t_2 - t_1) = -(\Delta M)_{dz} + 1/\rho_l \int_{t_1}^{t_2} (q_l + q_v) dt$$

where $\langle E \rangle$ is the mean evaporation rate from the soil surface over time $t_2 - t_1$ (m s^{-1}),

ρ_l is density of water (Mg m^{-3}),

q_l is liquid flux density at depth z (Mg m^{-2} s^{-1}),

q_v is vapor flux density at depth z (Mg m^{-2} s^{-1}), and

ΔM is the change in storage with depth (m), integrating 0 to z over time interval $(t_2 - t_1)$, and equals

$$\int_{t_1}^{t_2} \int_{0}^{z} \left(\frac{\delta \theta}{\delta t} \right) dz dt$$

where θ is volumetric water content (m^3 m^{-3}).

Jackson et al. (1974) concluded that the theory of Philip and de Vries (1957) best predicted the soil water flux under diurnal field conditions at intermediate water contents but that the isothermal model predicts better at high and low water contents. They calculated total water flux (J_w) as a function of temperature and water content gradients as follows:

$$J_w = -D(\theta)_l \nabla \theta - D(\theta)_v \nabla \theta - D(T)_l \nabla T - D(T)_v \nabla T - ki$$

where $D(T)_l$ is thermal liquid diffusivity,

$D(T)_v$ is thermal vapor diffusivity,

$\nabla \theta$ and ∇T are water and temperature gradients,

k is hydraulic conductivity, and

i is a unit vector.

The equation of Gardner (1959) did not take into account evaporation during the constant-rate stage. Ritchie (1972) modified Gardner's equation to include both stages and expressed cumulative evaporation (CE) as follows:

$$CE_1 = CE_0 \qquad \text{for } t < t_1 \text{ in which } CE_0 < CE_1$$
$$CE = CE_1 + \alpha(t - t_1)^{0.5} \qquad \text{for } t > t_1$$

where E_0 is the evaporativity and t_1 is the duration of the constant-rate stage. The value of α (slope of CE during the falling-rate stage versus $t^{0.5}$) was found to depend upon soil type (Ritchie, 1972) and E_0 (Jackson et al., 1976).

Stroosnijder and Kone (1982) modified Ritchie's (1972) equation for the falling-rate stage as follows:

$$CE_1 = CE_0 \qquad \text{for } t < t_1$$
$$CE = CE_1 + \alpha(t^{0.5} - t_1^{0.5}) \qquad \text{for } t > t_1$$

Instead of taking time as an independent variable, Boesten and Stroosnijder (1986) developed a parametric model by which evaporation is computed from the meteorological data only:

$$CE = CE_0 \qquad \text{for } CE_0 < \beta^2$$
$$CE = CE_0 \qquad \text{for } CE_0 = CE_1 = \beta^2$$
$$CE = \beta(CE_0)^{0.5} \qquad \text{for } CE_0 > \beta^2$$

where β is the evaporation parameter of soil (mm $mm^{-0.5}$) determined experimentally by plotting CE versus $(CE_0)^{0.5}$. In this model, CE depends on CE_0, not on time. This implies that a weight is attached to each day that is directly proportional to the evaporativity for the day. The parameter β was found to have little dependence on E_0.

An implicit assumption in the model is that $\int_0^t E_0(dt)$ does not depend upon E_0.

Heller (1968) presented mathematical analysis separately for the two stages of drying. In the first stage, soil is initially saturated fully with a liquid, which wets the grains. The vapor pressure at the top of the liquid is high enough for evaporation to proceed at a convenient rate but low enough that the temperature gradient produced by evaporative cooling has negligible effect on the drying rate. A small decrease in evaporation rate may occur as the liquid retreats to smaller pores and deeper crevices, which are less accessible to air currents. But so long as saturation at the surface is high enough for liquid flow, the liquid is replenished from below almost as fast as it evaporates. The second stage of drying is marked by the appearance of a bone-dry front (BDF), which is a fairly sharp transition zone between the lower part of the column, which contains some liquid saturation, and the part above it, which contains no free liquid. No liquid flow is presumed to occur between the surface and the BDF, and the critical rate is the rate of vapor transport upward through this region. This BDF is the site of evaporation during the second stage of drying and moves downward as evaporation proceeds.

Heller's (1968) analysis is purely theoretical and based on an approximately idealized system. The author used calcium fluorite, an inert, unconsolidated material in two sizes (80 and 140 mesh), instead of soil and chlorothen liquid instead of water. These factors tend to oversimplify the complex, heterogeneous system of soil, and his results may find limited applicability. The downward movement of the drying front was also simulated by van Keulen and Hillel (1974), who found it to move linearly with $t^{0.5}$.

Idso, Reginato, and Jackson (1979), while working on remote sensing of soil moisture and evaporation, developed a simple equation for estimating daily evaporation rate during all three stages of evaporation:

$$E_{i,ii,iii} = \left(\frac{3}{8} + \frac{5}{8}\beta\right)(S_n + 1.56L_n + 156)$$

where E is evaporation rate (cal cm^{-2} min^{-1}),
S_n is daily net solar radiation (cal cm^{-2} min^{-1}), and
L_n is net thermal radiation (cal cm^{-2} min^{-1}) and equals $R_a - R_s$
where R_a is extraterrestrial radiation (cal cm^{-2} min^{-1}) and is measured by the method given by Idso and Jackson (1969), and R_s is incoming radiation on the soil surface (cal cm^{-2} min^{-1}) and is estimated by the Stefan Boltzmann equation.

The quantity β is the partitioning factor:

$$\beta = \frac{\alpha_d - \alpha}{\alpha_d - \alpha_w}$$

where α_d is the albedo of dry soil,
α_w is the albedo of wet soil, and
α is the albedo for a particular day.

Jackson, Idso, and Reginato (1976) used the measured albedo to partition the total evaporating area into that evaporating at the energy-limited (potential) rate and that evaporating at the soil-limited rate to calculate actual evaporation rates during the transition from the energy-limited to the soil-limited phase. Since albedo is proportional to the surface water content, the day-to-day change in albedo indicates the fraction of the soil surface that is dry and is evaporating at the soil-limited rate. By denoting the partitioning factor β and by using a $t^{0.5}$ relation with a coefficient, C, for the soil-limited phase, the evaporation rate E_c for day n after start of drying was

$$(E_c)_n = \beta_n E_p + C\sum_{i=1}^{n}(\beta_{i-1} - \beta_i)(n - i + 1)^{-0.5} \tag{3.3}$$

where E_p is the energy-limited rate. Potential evaporation, E_p, was calculated with the formula of Priestley and Taylor (1972). By definition, $\beta_0 = 1$, since at time zero the soil is wet ($\alpha = \alpha_w$). An example will illustrate the calculations.

Suppose $n = 5$, $\beta_0 = 1$, $\beta_1 = 1$, $\beta_2 = 1$, $\beta_3 = 0.8$, $\beta_4 = 0.3$, $\beta_5 = 0.1$. Then, according to Equation 3.3, the evaporation rate at the end of day 5 will be

$$(E_c)_5 = 0.1E_p + C[0 + 0 + (1 - 0.8)(3)^{-0.5} + (0.8 - 0.3)(2)^{-0.5} + (0.3 - 0.1)(1)^{-0.5}]$$

Calculated evaporation rates were compared with lysimetrically determined rates. The method was found reliable for calculating evaporation rates during the transient phase (energy-limited to soil-limited) of soil drying.

Comparison of observed evaporation losses from silt loam, sandy loam, and loamy sand soils during the falling-rate stage with those predicted by different procedures showed that the procedure of Gardner (1959) was satisfactory for loamy sand (Jalota, 1990). The procedures of Gardner and Hillel (1962) and Jalota and Prihar (1987) were suitable for sandy loam and silt loam soils under evaporativities of 10.1 and 15.1 mm day^{-1}. Rowse's (1975) procedure underpredicted evaporation for all three soils and the two evaporativities (Figure 3.1). Inferentially, the assumptions made by Gardner (1959) to achieve the exact solution of the flow equation were valid only for coarse-textured soils, which dry quickly at the surface. Evidently, these assumptions did not hold for silt loams and sandy loams. Therefore, when predicting evaporation rates during the falling-rate stage from medium- and finer-textured soils, it is important to account for the effect of E_0 directly (such as by involving the drying pattern of the surface layer), as was done by Jalota and Prihar (1987), or indirectly (including surface water content changes in the average water content of the wetted length), as was done by Gardner and Hillel (1962).

Cumulative Evaporation-Time Relationships

Several researchers have attempted to quantify the relation between cumulative evaporation (CE) and time *(t)* from the solution of the flow equation (Gardner, 1959). Using a $t^{0.5}$ relation, Black, Gardner, and Thurtell (1969) computed evaporation from a bare Planfield sand (infinite length) within 5% variation of that measured with lysimeters. They observed a departure from a $t^{0.5}$ relation after rainfall because of the finite depth of wetting. An analysis by Rose (1969) predicted that evaporation proceeded linearly with the square root of time. Klute, Whisler, and Scott (1965) showed that for a finite length of wetting, the flux was initially proportional to $t^{0.5}$

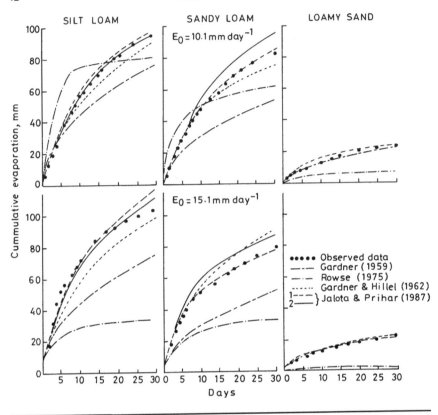

FIGURE 3.1. Comparison of observed and predicted (by different procedures) evaporation during the falling-rate stage from silt loam, sandy loam, and loamy sand soil columns under evaporativities of 10.1 and 15.1 mm day⁻¹ (Jalota, 1990)

but decreased sharply as the wetness at the bottom of the column was reduced. Similarly, Gardner (1974) observed that estimation of CE only as a function of $t^{0.5}$ is not sufficient, because a deviation was recorded just after a light rain. In the portion of that curve that departed from the square root of time, a correction was needed from the undisturbed-core data. From CE measurements on undisturbed cores of soil, a dimensionless curve was drawn relating water content loss to the square root of time divided by the amount of water available for evaporation. This system is applicable in an area with higher E_0 and rainfall. Kijne (1973) considered the evaporation losses from bare 0.13 m diameter and 0.35 m deep soil columns of fine sandy loam and clay loam soils during all three stages of evaporation under evaporativities of 3.1 and 9.4 mm day⁻¹ and found that CE during the first 4 days of drying after wetting was best approximated by a linear function of

TABLE **3.1** Best Fit of Cumulative Evaporation (CE) vs Time (t) during Different Periods of Drying under Two Evaporativities (E_0)

Soil type	E_0 (mm day^{-1})	Best-fit equation of CE*	Period for which relation holds (days)
Silt loam	15.6 ± 2.66	$14.1t^{0.72}$	0–7
		$24.2t^{0.43}$	7–15
		$30.13t^{0.38}$	15–30
		$-3.12 + 19.68t^{0.5}$	0–30
	4.6 ± 1.08	$5.3t^{0.80}$	0–15
		$20.29t^{0.36}$	15–30
		$-9.88 + 15.63t^{0.5}$	0–30
Sandy loam	15.6 ± 2.66	$12.8t^{0.71}$	0–7
		$27.6t^{0.29}$	7–15
		$30.1t^{0.26}$	15–30
	4.6 ± 1.08	$-6.6 + 11.1t^{0.5}$	0–30
Loamy sand	15.6 ± 2.66	$-0.007 + 5.8t^{0.5}$	0–30
	4.6 ± 1.08	$-1.12 + 5.6t^{0.5}$	0–30

Source: Jalota, 1984.

*CE is in millimeters and t is in days.

$t^{0.67}$ and during the next 15–20 days by a linear function of $t^{0.40}$. Jalota (1984), working with 0.95 m deep soil columns, found these exponents of time to vary with both soil type and E_0 (Table 3.1)

While fitting a $t^{0.5}$ relation to CE, a negative intercept was obtained. To account for these additional observations, Jalota, Prihar, and Gill (1988) modified the square root of time relation to include the constant-rate stage of evaporation. CE during the constant-rate stage and beyond obeyed the following relations:

$$CE = E_0 t \qquad \text{for } 0 < t < t_f \qquad (3.4a)$$

$$CE = Kt^{0.5} - \frac{K^2}{4E_0} \qquad \text{for } t > t_f \qquad (3.4b)$$

where E_0 is evaporativity, K is the regression coefficient of CE versus $t^{0.5}$, and t_f is duration (in days) of the constant-rate stage, which was shown to be related to K and E_0 through Equation 3.4c:

$$t_f = \frac{K^2}{4E_0^2} \qquad (3.4c)$$

Use of the modified square root relation, despite its simplicity, remained limited because the regression coefficient of CE versus $t^{0.5}$ was affected by the

TABLE **3.2**   Regression Constants for the (Observed) Cumulative Evaporation and Square Root of Time Relation (CE = $a + Kt^{0.5}$) for Different Evaporativities (E_0) and Soil Textures (CE includes Constant-Rate Stage Evaporation)

E_0 (mm day^{-1})	Silt loam			Sandy loam			Loamy sand		
	a	K	r^{2*}	a	K	r^2	a	K	r^2
2.5	−7.7	9.36	.99	−10.6	10.60	.90	−0.6	4.07	.99
3.6	−5.9	11.07	.99	−5.3	11.01	.98	−1.0	3.22	.99
4.0	−5.9	11.92	.98	—	—	—	1.9	3.76	.99
5.2	−2.6	14.01	.98	−5.7	13.11	.99	—	—	—
6.0	−5.9	16.3	.99	−5.3	14.43	.98	−0.7	4.50	.99
10.4	−8.4	21.09	.98	−4.4	17.56	.99	−1.3	4.99	.99
13.0	−8.3	22.59	.98	−1.8	15.74	.99	2.0	3.76	.99
15.1	−4.1	23.00	.99	−1.9	13.82	.99	−2.4	3.64	.99

Source: Jalota, 1994.
*r^2 = coefficient of determination.

water transmission properties of soil and by E_0 (Jackson, Idso, and Reginato, 1976; Hall and Dancette, 1978) under field conditions. To facilitate the use of this equation under diverse situations, Jalota (1994) determined K values for three soils under a variety of evaporativities (Table 3.2) and found the coefficient thus obtained to be related to soil water diffusivity ($\bar{D}$) and E_0 as in Equation 3.5:

$$K = 2.908 + 0.00298\bar{D} + 1.032E_0 + 0.0006E_0\bar{D} - 0.095E_0^2 \qquad (3.5)$$
$$(r^2 = .96, n = 22)$$

where K is an evaporation parameter (mm day$^{-0.5}$), $\bar{D}$ is mean weighted diffusivity (mm^2 day^{-1}), and E_0 is evaporativity (mm day^{-1}).

This model was tested on data of Willis and Bond (1971), Black, Gardner, and Thurtell (1969), and Singh, Oswal, and Jagan Nath (1985). The observed CE values agreed closely (Figure 3.2) with those computed for their experimental conditions using Equations 3.4b and 3.5 (Jalota, 1994).

Numerical Approach

The second approach for predicting the magnitude and rate of evaporation is to use numerical techniques. Two main systems that ordinarily interact during

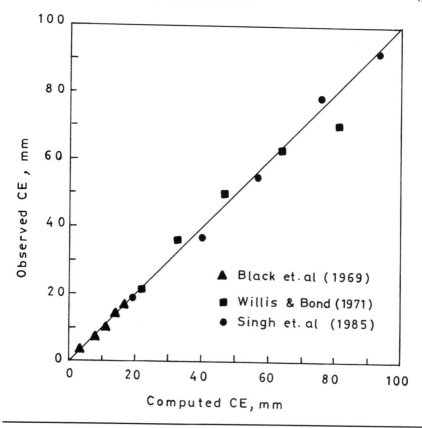

FIGURE 3.2.   Comparison between the CE values observed by Black, Gardner, and Thurtell (1969), Willis and Bond (1971), and Singh, Oswal, and Nath (1985) and those computed by Equations 3.4b and 3.5 for their respective experimental conditions (Jalota, 1994)

the evaporation process are (1) the atmospheric system, which provides energy to meet the latent heat of evaporation, and (2) the soil system, which continuously supplies water to the evaporation site.

The Atmospheric System

In the atmospheric system, the energy balance at the soil surface determines the E_0. Methods available for computing E_0 can be grouped into three categories: (1) aerodynamic methods, (2) energy balance methods, and (3) a combination approach.

Aerodynamic Methods. In the aerodynamic approach wind profiles and vapor pressure differences above the ground surface are characterized. Three equations are fundamental to this approach. The first is the Thornthwaite-Holzman (1942) equation:

$$q_a = \frac{\rho_a \varepsilon k^2 (\overline{U}_2 - \overline{U}_1)(\overline{e}_1 - \overline{e}_2)}{P_a [\ln(z_2 / z_1)]^2}$$

where q_a is evaporative flux (mm day^{-1}),
ρ_a is density of air (Mg m^{-3}),
k is Karman's constant (= 0.41),
ε is the ratio of the weight of water vapor to that of air,
$\overline{U}_2 - \overline{U}_1$ is the difference in wind speed (m s^{-1}) measured over two observational heights, z_2 and z_1,
$\overline{e}_1$ and $\overline{e}_2$ are vapor pressures at z_1 and z_2, respectively (MPa), and
P_a is atmospheric pressure (MPa).

The Penman (1956) equations are as follows:

$$E = 0.35(e_0^0 - e_d)[1 + (u_2/100)] \tag{3.6a}$$
$$E = 0.35(e_0^0 - e_d)[0.5 + (u_2/100)] \tag{3.6b}$$

where E is evaporation rate (mm day^{-1}),
u_2 is average wind speed at 2 m above the drying surface (mi day^{-1}),
e_d is the actual vapor pressure (mm Hg) in the air at observational height d (2 m) above the surface, and
e_0^0 is the saturated vapor pressure (mm Hg) at the temperature of the surface T_0.

Energy Balance Method. In the energy balance approach, all other sources of and sinks for energy except evaporation are measured. The energy balance at the soil surface can be written as

$$R_n = G + q_n + \text{LE}$$

where R_n is net radiation flux (W m^{-2}),
G is soil heat flux (W m^{-2}),
q_n is sensible heat flux (W m^{-2}), and
LE is latent heat flux associated with evaporation (W m^{-2}).

When the soil is fully saturated and the wind is calm, a greater proportion of the solar radiation goes into evaporating water, and hence, the values of G and q_n are small. A rough estimate of LE can be made directly from the R_n. On a bare surface, the net radiation is measured directly by net radiometers or is estimated by Equation 3.7, suggested by Penman (1956):

$$R_n = R_a(1 - r)(0.18 + 0.55n/N) - \sigma T^4(0.56 - 0.92\bar{e}_2^{0.5})(0.10 + 0.09n/N) \qquad (3.7)$$

where R_a is mean extraterrestrial radiation (mm water day^{-1}),
r is the reflection coefficient, or albedo (0.05 for a water surface),
n is actual duration of bright sunshine (hr),
N is maximum possible duration of bright sunshine for a cloudless sky (hr),
σ is the Boltzmann constant (mm day^{-1} K^{-4}),
T is mean air temperature (K), and
$\bar{e}_2$ is actual vapor pressure at 2 m height (mm Hg).

Net radiation is positive during the day and negative during the night. Since energy is partitioned among several processes such as heating the soil and sensible heat, R_n may have positive or negative values depending upon the prevailing conditions. To overcome these difficulties, Bowen (1926) suggested the use of the ratio of q_n to LE, which later came to be known as the Bowen ratio, β. In equation form it is expressed as

$$\beta = \frac{q_n}{\text{LE}} = r\frac{\Delta T}{\Delta e}$$

and hence the evaporation flux, E, can be expressed as

$$E = \frac{R_n - G}{L(1 + \beta)}$$

where ΔT is $T_2 - T_1$ and Δe is $e_2 - e_1$ measured over $\Delta z = z_2 - z_1$ (the observational heights) and

$$r = P_a C_p / \varepsilon L$$

is the psychrometer constant (Pa °C^{-1}), where
P_a is atmospheric pressure (Pa),
C_p is specific heat of air at constant pressure (J kg^{-1} °C^{-1}),
ε is the ratio of the molecular weight of water to that of air (= 0.622), and
L is latent heat of evaporation (J m^{-3}).

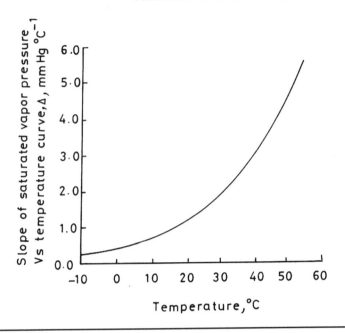

FIGURE 3.3.   Saturated vapor pressure as a function of temperature

This equation predicts E satisfactorily when evaporation flux is larger than heat flux.

Combination Method. In this method evaporation is determined from the measurements made at a single height in the air assuming that vapor pressure at a sufficiently wet surface equals the saturated vapor pressure at the temperature of the surface, T_0. This eliminates the need for measuring vapor pressure at the surface. The equation of the combination method (van Bavel, 1966) is as follows:

$$LE = \frac{\Delta}{\Delta + r} R_n + \frac{r}{\Delta + r} LE_a$$

The value of Δ is computed as the slope of the saturated vapor pressure versus T curve (Figure 3.3) for temperature measured at the selected height. The value of Δ/r for temperature measured at the preselected height is taken from Table 3.3, and E_a is calculated from Penman's equations for E (Equations 3.6a and 3.6b).

TABLE 3.3 Δ/r (Dimensionless) Values for T in °C at 1000 mb Pressure

T	Δ/r	T	Δ/r
0.0	0.67	24.0	2.64
1.0	0.72	26.0	2.92
2.0	0.76	28.0	3.23
3.0	0.81	30.0	3.57
4.0	0.86	32.0	3.93
5.0	0.92	34.0	4.32
6.0	0.97	36.0	4.72
7.0	1.03	38.0	5.20
8.0	1.10	40.0	5.70
9.0	1.16	42.0	6.23
10.0	1.23	44.0	6.80
11.0	1.30	46.0	7.41
12.0	1.38	48.0	8.07
14.0	1.55	50.0	8.77
16.0	1.73	52.0	9.52
18.0	1.93	54.0	10.3
20.0	2.14	56.0	11.3
22.0	2.38	58.0	12.1

Monteith's (1965) version of Penman's (1948) equation is also used to calculate LE:

$$LE = \frac{S(R_n - G)\rho_a C_p (\delta e / r_a)}{S + r} \tag{3.8}$$

where R_n is net radiation (W m^{-2}),
ρ_a is density of air (Mg m^{-3}),
C_p is specific heat of air at constant pressure (J Mg^{-1} °C^{-1}),
δe is the vapor pressure deficit in air (Pa),
S is the slope of the saturation vapor pressure curve (Pa °C^{-1}) at air temperature,
r is the psychrometer constant (Pa °C^{-1}), and
r_a is aerodynamic resistance (s m^{-1}).

The aerodynamic resistance depends upon the wind speed and roughness coefficient. It is calculated with the following equation (van Bavel and Hillel, 1976):

$$r_a = \frac{[\ln(z / z_0)]^2}{0.16U}$$

where z_0 is roughness length (m),
U is the wind speed (m s^{-1}), and
z is the reference level at which U is measured (2 m).

The Soil System

Equation 3.1 is solved numerically according to the set of flow conditions subject to certain boundary conditions at the surface and bottom of drying soil columns.

The boundary conditions at the soil surface and atmospheric interface were represented by Philip (1957) and Reynolds and Walker (1984) as follows:

$$q_s = -D(\theta_s)\frac{\delta\theta}{\delta z}\bigg|_{z=0} = q_a = \frac{-\rho_a C_p RH[e_s(T_s)-e_a]}{\tau r_a} \tag{3.9}$$

where q_s and q_a are evaporative flux across the soil surface as defined by soil moisture flow and atmospheric water flow, respectively (m s^{-1}),
θ_s is soil moisture at the soil surface,
$e_s(T_s)$ is saturated vapor pressure of the air at the soil surface (Pa),
e_a is atmospheric vapor pressure (Pa),
τ is a psychrometer constant (Pa °C^{-1}),
r_a is aerodynamic resistance (s m^{-1}), and
RH is relative humidity of air at soil surface temperature (T_s). It is computed as a function of temperature and volumetric water content (θ_s) as follows:

$$RH = \frac{\exp[g(\theta_s)]}{RT_s}$$

where R and g are gas and gravitational constants, respectively.

Assuming that the water potential (h) at the soil surface is in equilibrium with the atmospheric vapor potential, the quantity $h(0,t)$ is estimated daily using relative humidity and temperature data with Equation 3.10:

$$h(0,t) = \frac{RT(t)}{gM}\ln f(t) \tag{3.10}$$

where h is the soil water potential (m),
R is the gas constant (J mole^{-1} K^{-1}),
T is absolute temperature (K),

g is acceleration due to gravity (m s^{-2}),

M is the molecular weight of water (kg mole^{-1}), and

$f(t)$ is the relative humidity of air as a function of time.

If one knows h at the soil surface at any time, the water content of that layer at that time can be estimated from the relation

$$\theta = a \exp(-bh)$$

where θ is water content (m^3 m^{-3}), h is soil water potential (Pa), and a and b are empirical constants obtained by curve fitting to soil water characteristic data.

The boundary conditions at the bottom of the soil column are

$$q_b = D(\theta_b)\left.\frac{\delta\theta}{\delta z}\right|_{z=L} = 0 \tag{3.11}$$

where q_b is evaporative flux at this boundary, θ_b is volumetric moisture content at the bottom, and L is the length of the column.

Knowledge of the initial soil moisture profile is a prerequisite for running the model. The most important part of Equations 3.1, 3.9, and 3.11 is specifications of flow functions: $D(\theta)$ and $h(\theta)$ versus the independent variable θ. For $D(\theta)$ the equation used is of the type

$$D(\theta) = A \exp(B\theta)$$

in which A and B are positive constants determined from a curve fitted to experimental values of $D(\theta)$ versus θ. For $h(\theta)$ the selected function is of the form

$$h = a \exp(b\theta)$$

where a and b are constants. Any other function suitable for the given type of soil can also be used.

Numerical Solution Techniques

The nonlinear partial equations (Equations 3.1, 3.9, and 3.11) are solved numerically with a finite-difference scheme. In the first step, the soil profile length (L) is divided into N number of sublayers (N is chosen arbitrarily) of

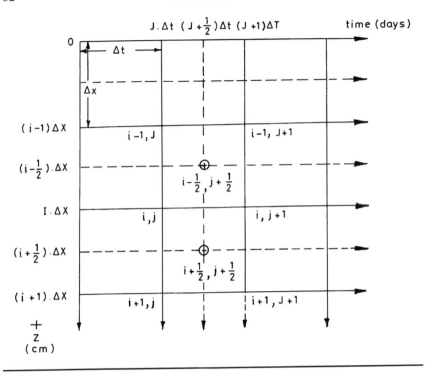

FIGURE 3.4. Model discretization

constant thickness x ($x = L/N$) from the soil surface ($N = 0$) to the bottom of the profile. A time interval Δt is chosen that depends upon the stability of the process. Then Equations 3.1, 3.9, and 3.11 are written in terms of Crank-Nicholson's (1947) implicit scheme (Figure 3.4).

The partial-differential terms are approximated as follows:

$$\frac{\delta\theta}{\delta t} = \frac{1}{\Delta z}\{[D(\theta)_{i+1/2,j+1/2}\left(\frac{\delta\theta}{\delta z}\right)_{i+1/2,j+1/2}] - [D(\theta)_{i-1/2,j+1/2}\left(\frac{\delta\theta}{\delta z}\right)_{i-1/2,j+1/2}]\}$$

where subscripts i and j are space and time variables, respectively; Δt is the time interval; and Δz is the depth increment.

In the second step, $D(\theta)$ at $\theta_{i+1/2,j+1/2}$ and $\theta_{i-1/2,j+1/2}$ (i.e., volumetric water content at half space and half time levels) are calculated:

$$\frac{\delta\theta}{\delta t} = \frac{\theta_{i,j+1} - \theta_{i,j}}{\Delta t}$$

$$\left(\frac{\delta\theta}{\delta z}\right)_{i+1/2,j+1/2} = \frac{1}{2\Delta z}[(\theta_{i+1,j+1} + \theta_{i+1,j}) - (\theta_{i,j+1} + \theta_{i,j})]$$

$$\left(\frac{\delta\theta}{\delta z}\right)_{i-1/2,j+1/2} = \frac{1}{2\Delta z}[(\theta_{i,j+1} + \theta_{i,j}) - (\theta_{i-1,j+1} + \theta_{i-1,j})]$$

$$D(\theta)_{i+1/2,j+1/2} = D[(3/4)(\theta_{i+1,j} + \theta_{i,j}) - (1/4)(\theta_{i+1,j-1} + \theta_{i,j-1})]$$

$$D(\theta)_{i-1/2,j+1/2} = D[(3/4)(\theta_{i,j} + \theta_{i-1,j}) - (1/4)(\theta_{i,j-1} + \theta_{i-1,j-1})]$$

Let

$$D[3/4(\theta_{i+1,j} + \theta_{i,j}) - 1/4(\theta_{i+1,j-1} + \theta_{i,j-1})] = B$$

and

$$D[3/4(\theta_{i,j} + \theta_{i-1,j}) - 1/4(\theta_{i,j-1} + \theta_{i-1,j-1})] = F$$

The known values of θ in Figure 3.4 are $\theta_{i-1,j}$, $\theta_{i,j}$, and $\theta_{i+1,j}$. The unknown are $\theta_{i-1,j+1}$, $\theta_{i,j+1}$, $\theta_{i+1,j+1}$, $\theta_{i-1/2,j+1/2}$, and $\theta_{i+1/2,j+1/2}$. The values $\theta_{i+1/2,j}$ and $\theta_{i-1/2,j}$ are calculated as the arithmetic average of values at nodes $i - 1,j$; i,j; and $i + 1,j$ as follows:

$$\theta_{i+1/2,j} = (\theta_{i,j} + \theta_{i+1,j})/2$$

$$\theta_{i-1/2,j} = (\theta_{i,j} + \theta_{i-1,j})/2$$

The value $\theta_{i+1/2,j}$ at half time $j + 1/2$ is estimated by backward Taylor series projection (von Rosenberg, 1969).

The surface boundary is represented by

$$q_s = D(\theta_s)\frac{\theta_{s-1,j} - \theta_{s+1,j}}{2\Delta z} = \frac{\rho_a C_{pa}[RHe_s(T_s) - e_a]}{\tau r_a} \tag{3.12}$$

where θ_{s-1}, θ_{s+1} and θ_s designate the volumetric soil water content at fictitious nodes immediately above the surface, immediately below the surface, and at the soil surface, respectively.

Equation 3.12 is solved together with Equation 3.1, which is written in Crank-Nicholson terms of $\theta_{s,j+1}$, by a regular false iterative method. The convergence criterion (CR) in this numerical integration is

$$CR = \left| \theta_{s,j+1}^{K+1} - \theta_{s,j+1}^{K} \right| = 10^{-4}\,\text{m}^3\,\text{m}^{-2}\,\text{s}^{-1}$$

where K is the number of iterations.

In the third step, computation is done at each j loop, allowing $\theta_{i,j+1}$ of the Crank-Nicholson-transformed equation (Equation 3.1) to be obtained.

In the fourth step, actual evaporation is computed from integration of the continuity equation

$$\left(\frac{\delta q}{\delta z} \right)_t = -\left(\frac{\delta \theta}{\delta t} \right)_z$$

which can be written in integral form as follows:

$$q(z) = -\int_{z=L}^{z} \frac{\delta \theta}{\delta t}\,dz$$

which provides the flux values at each z level for each time integration and, thus, the flux value at surface $z = 0$. The integration is next achieved by a trapezoidal rule, and actual evaporation at each time step is given by $q\,(z = 0)$. Reynolds and Walker (1984) reported that running a specific and cumulative mass balance check at each time step improves the efficiency of the program. For that, atmosphere-based (q_a) and soil-based (q_s) evaporation is calculated between the two times. The specific mass balance check has the form

$$RD_1 \leq \left| \frac{q_a - q_s}{S} \right| \leq RD_2$$

where S is q_a or q_s, whichever is numerically smaller, RD_1 is specific minimum discrepancy in flux estimate (= 0.002 mm hr^{-1}) for all simulations, and RD_2 is specific maximum discrepancy in flux estimate (= 0.3 mm hr^{-1}) for all simulations. If the resultant is less than RD_1, the next time step is made 2% larger than the present. And if the resultant is greater than RD_2, the present time step is decreased by 2%, and all calculations are repeated. The cumulative mass balance check consists of the determination of cumulative mass transfer across the surface boundary according to atmosphere-based and soil-based flux calculations. Reynolds and Walker (1984) used soil surface temperature as a function of time, $T_s(t)$, as the forcing function in the surface boundary, which is both feasible and practical. In addition, the use of measured $T_s(t)$ precludes the necessity of conducting a complex energy balance for the soil surface, as done by Flerchinger and Saxton (1989).

Jalota and Prihar (1987) incorporated the gravity term into the flow equation (Equation 3.1), which resulted in the following equation:

$$\frac{\delta\theta}{\delta t} = \frac{\delta}{\delta z}\left[D(\theta)\frac{\delta\theta}{\delta z}\right] - \frac{\delta}{\delta z}K(\theta) \tag{3.13}$$

It was solved numerically in two ways. In one case (I) the evaporation rate from the soil was equated with the flux density through the (shallow) plane at 2.5 cm below the surface. In the second procedure (II) it was assumed that drying of the surface layer integrates the effects of E_0, initial wetness, and the $D(\theta)$ or $K(\theta)$ relations of the soil. Equation 3.13 was solved by the finite-difference technique for the following initial and boundary conditions:

$$\theta(z,0) = \theta_i(z) \qquad\qquad\qquad \text{(initial condition)}$$

$$\theta(\Delta z_1,t) = \theta_1(t) \qquad\qquad\qquad \text{(upper boundary condition)}$$

$$\left(\frac{\delta\theta}{\delta z}\right)_{z=L} = 0 \qquad\qquad\qquad \text{(lower boundary condition)}$$

where θ_i is the initial water content, and θ_1 is the water content of the first soil node, Δz_1.

The observed CE agreed well with that calculated by both procedures in a sandy loam soil (Figure 3.5). A numerical approach to simulating water and temperature profiles for a bare soil has also been used successfully in the CONSERVB Model by Lascano and van Bavel (1983, 1986).

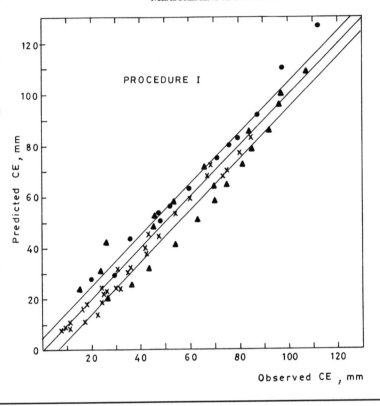

FIGURE 3.5.   Comparison of the observed CE from sandy loam soil with that computed from Equation 3.13 using procedures I and II under three evaporativities (Jalota and Prihar, 1987)

Evaporation from Tilled Bare Soil

Analytical Approach

Methods of predicting evaporation from tilled soil range from ordinary regression equations to process-based soil-water-atmosphere flow equations. Hanks (1958) and Hanks and Woodruff (1958) attempted to model the effect of tillage depth and porosity of the tilled layer on evaporation using surfactant-treated soil, assuming instant drying of the tilled soil mass. The vapor flux across the (simulated) tilled layer was computed by assuming the lower boundary of the layer as the locus of phase change using the following equation (Hanks, 1958):

$$q = -\alpha \phi \frac{DM}{RT} \frac{P}{P - P_v} \frac{dP_v}{dx}$$

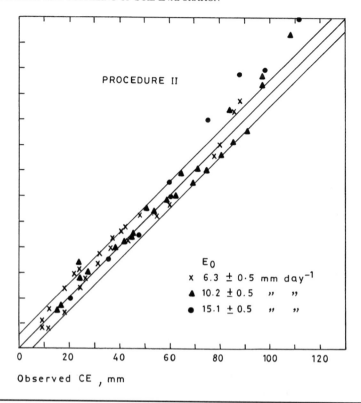

FIGURE 3.5. Continued

where q is water flow rate (kg m^{-2} s^{-1}),
D is the diffusion coefficient of water vapor into still air (m^2 s^{-1}),
P is total pressure (Pa),
P_v is partial pressure of water vapor (Pa),
R is the gas constant (J K^{-1} mole^{-1}),
T is temperature (K),
M is molecular weight of water (kg mole^{-1}),
α is a tortuosity factor,
ϕ is the volume fraction of air-filled voids, and
x is depth of soil mulch (m).

The diffusion coefficient D is a function of temperature and pressure. Krischer and Rohnalter (1940) showed the relation to be of the form

$$D = D_0 \frac{P_0}{P} \frac{T^{2.3}}{T_0^{2.3}}$$

where D_0, P_0, and T_0 are appropriate values for standard conditions, and P and T are any pressure and temperature.

The value of T was taken as the temperature at the bottom of any layer. Hillel, van Bavel, and Talpaz (1975) developed an equation by combining the well-known combination equation of van Bavel (1966) with that of Fuchs and Tanner (1967), assuming that air temperature over the simulated tilled and untilled soil was the same and reflectivity and heat flux were unaffected by the soil mulch. The resultant equation (Equation 3.14) involved a diffusion resistance parameter of the soil mulch that depended on the porosity, depth, and tortuosity of the soil mulch and was used to simulate the dynamics of water storage in the profile:

$$\frac{E_m}{E_0} = \frac{\varepsilon + 1}{\varepsilon + 1 + r_m/r_a} \tag{3.14}$$

where E_m is the evaporation rate from a mulch-covered surface (kg m^{-2} s^{-1}), E_0 is evaporativity (kg m^{-2} s^{-1}),
$\varepsilon = \Delta/r$, dimensionless, where $\Delta = dv/dT$ = slope of saturated vapor density versus temperature curve (kg m^{-3} degree^{-1}), and r is the psychrometer constant (kg m^{-3} degree^{-1}),
$r_m = M/D\phi\alpha$, where M is mulch thickness (m), D is diffusivity of bulk air (m^2 s^{-1}), ϕ is mulch porosity (fraction), and α is tortuosity, dimensionless, and r_a is aerodynamic resistance to vapor flow between the evaporating surface and elevation z (s m^{-1}).

Tselishcheva (1965) used the equation given by Budagoveskly (1962), who showed that the ratio of evaporation from the wet surface of untilled soil (E_0) to evaporation from dried tilled soil (E) was related to the thickness of the soil mulch Z_n (cm):

$$\frac{E_0}{E} = 1 + BDZ_n \tag{3.15}$$

where B in Equation 3.15 is the value depending upon free porosity, the coefficient of heat conductivity, air temperature, and other values according to equality:

$$B = (Pi + 0.622\rho L\phi'\varepsilon)C_p/(0.622L\phi' + PC_p)i\varepsilon$$

where P is atmospheric pressure (millibars),
i is the coefficient of heat conductivity (cal cm^{-1} s^{-1} $°C^{-1}$),
ρ is air density (g cm^{-3}),
L is latent heat of evaporation (cal g^{-1}),
C_p is specific heat of air at constant pressure (cal g^{-1} $°C^{-1}$),
ϕ' is the derivative of maximum vapor pressure tension when $T = TB$ (TB is the air temperature of the surrounding air),
ε depends upon the coefficient of molecular diffusion of water vapor and free porosity of the soil, and
D is the coefficient of exchange between the soil surface and the surrounding air.

Acharya and Prihar (1969) tested Equation 3.15 and concluded that it did not hold good for all conditions of porosity and wind velocity. It was found valid for porosities up to 0.62 when wind velocity was less than 6 km hr^{-1}, and for mulch porosities less than 0.59 when wind velocity was 12 km hr^{-1} or less. Within these limits the following multiple-regression equation described the integrated constant BD in terms of mulch porosity (x_1) and wind velocity (x_2):

$$BD = a + bx_1 + cx_2$$

where a, b, and c are empirical constants.

The BD values were obtained using E_0/E values from the same wind velocity. Within the limits given above, Equation 3.15 permits computation of evaporation through dry soil mulch from the porosity and thickness of mulch, wind velocity, and E_0. Gardner and Fireman (1958) gave the following empirical relationship for evaporation from soil covered with dry sand mulch:

$$E = D_m \frac{P_1 - P_2}{L}$$

where E is evaporation rate,
L is the thickness of sand mulch,
P_1 is the saturated vapor pressure of soil,
P_2 is the vapor pressure at the soil surface, and
D_m is the molecular diffusion coefficient.

Evaporativity was either accounted for in the computation of vapor pressure at the surface (Hanks, 1958) or included as such (Acharya and Prihar, 1969; Budagoveskly, 1962).

TABLE 3.4 Evaporation Parameter K^* (in mm Day$^{-0.5}$) as Influenced by Evaporativity (E_0) and Depth of Tillage

		Silt loam			Sandy loam		
E_0 (mm day^{-1})	Depth of tillage (cm)	K	a	$r^{2\dagger}$	K	a	r^2
2.6	5	5.4	−4.7	.99	2.0	−0.9	.99
3.6	2	—	—	—	4.5	−3.1	.97
3.6	5	—	—	—	3.3	−3.0	.97
5.2	2	6.2	−3.4	.98	4.4	−2.9	.96
5.2	5	6.5	−2.5	.98	3.9	−1.4	.98
10.4	2	9.4	−9.0	.93	7.7	−4.3	.99
10.4	5	4.5	−0.7	.99	4.9	−2.5	.97
13.0	2	10.3	−7.7	.95	8.1	−6.1	.96
13.0	5	8.7	−6.2	.97	6.5	−5.1	.93
15.1	2	12.3	−9.8	.95	10.7	−7.8	.95
15.1	5	11.1	−7.3	.97	7.1	−4.2	.95

Source: Jalota, 1995.
*For the relation CE = $a + Kt^{0.5}$.
†r^2 coefficient of determination.

Gill and Prihar (1983) modeled evaporation from tilled soil as a linear function of the square root of time. They assumed that the effect of cultivation was similar to that of intensely high E_0 because both conditions hasten the dry-layer formation. Tillage depth and E_0 effects were reflected in the regression coefficients. Jalota (1995) showed that evaporation parameter K in the relationship CE = $a + Kt^{0.5}$ is also influenced by soil type (Table 3.4). Values of K (mm day$^{-0.5}$) for untilled (Table 3.2) and tilled soil (Table 3.4) multiply regressed over tillage depth (TD, in cm) and E_0 (mm day^{-1}) yielded the following equations for silt loam and sandy loam soils for E_0 ranging between 2.5 and 15.1 mm day^{-1} and for tillage depth between 0 and 5 cm.

Silt loam:

$$K = 9.90 + 0.89E_0 - 6.69\text{TD} + 1.05\text{TD}^2 - 0.0012\text{TD}E_0 \quad (r^2 = .93, n = 18) \qquad \textbf{(3.16a)}$$

Sandy loam:

$$K = 10.82 + 0.38E_0 - 4.61\text{TD} - 0.57\text{TD}^2 + 0.000064\text{TD}E_0 \quad (r^2 = .97, n = 18) \qquad \textbf{(3.16b)}$$

CE values computed with Equations 3.4a and 3.4b using the K values determined from Equations 3.16a and 3.16b were tested on the data of Willis and Bond (1971) for fine sandy loam soils and on the data of Gill (1992) for

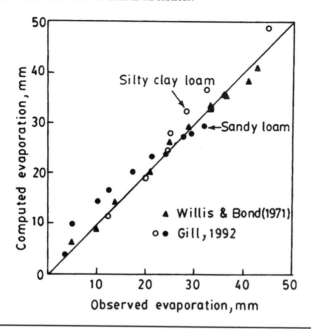

FIGURE 3.6.   Comparison between computed CE using Equations 3.4a and 3.4b (involving the estimated evaporation parameter from Equations 3.16a and 3.16b) and observed data of Willis and Bond (1971) and Gill (1992) (Jalota, 1995)

silty clay loam and sandy loam soils. The observed and computed CE values (Figure 3.6) exhibited reasonable agreement.

Numerical Approach

As mentioned in Chapter 2, tillage alters the heat flux through changed roughness, which changes the area in contact with the atmosphere and traps radiation. Under such conditions, drying becomes nonisothermal. Therefore, using the isothermal model might introduce error in the estimation of evaporation. Hanks, Gardner, and Fairbourn (1967) and Hadas (1975) estimated this error to be as high as 20% for homogeneous soil and conjectured that it could be still higher for layered soils, especially when there are thermal fluctuations. Hillel and Talpaz (1977) used arbitrary soil properties and weather patterns to simulate soil water dynamics for isothermal conditions and predicted evaporation from variously constituted nonhomogeneous profiles. But these studies did not take into account interfacial impedance to water flow, which is very important in layered soils (Hillel and Hadas, 1972). Under thermal fluctuations, vapor flow contributes significantly at intermediate water contents (Jackson et al., 1974). Hammel, Papendick, and Campbell (1981) did account for vapor flow while solving the flow equation for combined moisture and temperature gradients. They solved the equation of Philip and de Vries (1957) given below (Equation 3.17) using the Crank-Nicholson time-centered finite-difference scheme (Crank, 1956) and predicted changes in moisture content at different depths in tilled and untilled soil:

$$\frac{\delta\theta}{\delta t} = \frac{\delta}{\delta z}\left\{[D(\theta)_{\text{liq}} + D(\theta)_{\text{vap}}]\frac{\delta\theta}{\delta z} + D(T)_{\text{vap}}\frac{\delta T}{\delta z}\right\} \tag{3.17}$$

where $D(\theta)_{\text{liq}}$ is the isothermal liquid water diffusivity (m^2 s^{-1}), $D(\theta)_{\text{vap}}$ is the isothermal vapor diffusivity (m^2 s^{-1}), $D(T)_{\text{vap}}$ is the nonisothermal vapor diffusivity (m^2 s^{-1}), T is the temperature (°C), and z is a vertical coordinate (taken positive downward) (m).

Although a number of mechanistic models based on water and heat flow theory (van Keulen and van Beek, 1971; Roesma, 1975; Linden, 1982; Hammel, Papendick, and Campbell, 1981; Camillo, Gurney, and Schmugge, 1983) have been developed, none of these models considers the phenomenon of an evaporation front moving downward with time (as described in Chapter 2). For any mechanistic model, input data should include the downward movement of the locale of evaporation, the soil temperature wave in tilled and untilled soil, and wind penetration as affected by clod size.

Residue Effects and the Interaction of Residue with Tillage

Analytical Approach

Prihar, Jalota, and Steiner (1996) used Ritchie's (1972) model and the modified square root of time relation proposed by Jalota, Prihar, and Gill (1988) to predict evaporation from residue-covered and tilled soils. Evaporation was computed in two steps. Since the residue mulch reduces the rate of evaporation during the constant-rate stage, it is imperative to first quantize the evaporation rate from mulch-covered soil in terms of mulch rate (the extent and depth of coverage of the mulch) and evaporativity. Further analysis was carried out as if the mulched soil was drying as bare soil under the reduced rate of constant-rate stage evaporativity. For the simple parametric model of Ritchie, the CE values were computed as follows:

1. The energy-limited rate of evaporation from mulched soil (E_m) was obtained from E_0 and mulch rate using the empirical constants of Table 3.5 for each soil.
2. The duration of the energy-limited stage of evaporation in different soils was determined from the equations of Figure 2.13.

3. The evaporation during the energy-limited stage, U, was computed as the product of E_m and the duration of the stage.

4. Evaporation during the soil-limited stage at a given time was computed as the product of α (Table 3.6) and $(t - t_1)^{0.5}$.

5. CE at a given time was the sum of U and evaporation during the soil-limited stage.

The computed and observed values of CE agreed closely for the loamy sand and clay loam soils under both evaporativities (Figure 3.7) and for all

TABLE 3.5. Ratio of Energy-Limited Evaporation Rate from Mulched Pots to Free Water Evaporation (E_m/E_0) as a Function of Residue Rate (RES) and Evaporativity (E_0)

Soil type	a	b	c	r^2
Silt loam	1.061	−0.218	−0.0221	.95
Sandy loam	1.040	−0.225	−0.0074	.99
Loamy sand	1.223	−0.224	−0.0388	.90

Source: Prihar, Jalota, and Steiner, 1996.
Note: $E_m/E_0 = a \exp[-(b\text{RES} + cE_0)]$. RES is in Mg ha^{-1}.

TABLE 3.6 Effect of Tillage, Mulch Rate (MR), Residue Management, E_0, and Soil Type on the Parameter α in Ritchie's Model: $CE = U + \alpha(t - t_1)^{0.5}$

	Mulch rate (Mg ha^{-1})					
	$E_0 = 10.0$ mm day^{-1}			$E_0 = 2.5$ mm day^{-1}		
Treatments	0	3	6	0	3	6
	α (mm day$^{-0.5}$)					
	Silt loam					
No tillage	11.2	13.4	11.0	6.5	6.0	5.8
Tillage	—	7.3	6.6	—	3.5	3.8
Undercutting	—	8.7	7.6	—	3.6	3.0
Mixed	—	5.5	5.0	—	3.5	2.6
	Sandy loam					
No tillage	11.7	11.9	10.2	5.2	5.2	5.4
Tillage	—	11.5	6.0	—	2.9	2.5
Undercutting	—	9.0	9.9	—	3.5	2.6
Mixed	—	7.2	5.3	—	2.3	2.7
	Loamy sand					
No tillage	5.1	5.2	4.4	3.3	2.9	2.3
Tillage	—	4.1	3.4	—	2.1	1.5
Undercutting	—	3.6	3.3	—	1.8	1.8
Mixed	—	2.9	5.3	—	1.8	1.3

Source: Prihar, Jalota, and Steiner, 1996.

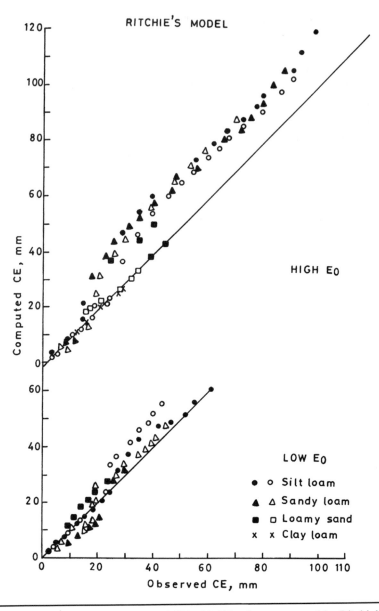

FIGURE **3.7.** Comparison of observed CE with CE values computed by Ritchie's (1972) and Jalota, Prihar, and Gill's (1988) procedures for 3 Mg ha^{-1} (*solid symbols*) and 6 Mg ha^{-1} (*hollow symbols*) mulched silt loam, sandy loam, loamy sand, and clay loam soils under evaporativities of 2.5 and 10.0 mm day^{-1} (Prihar, Jalota, and Steiner, 1996)

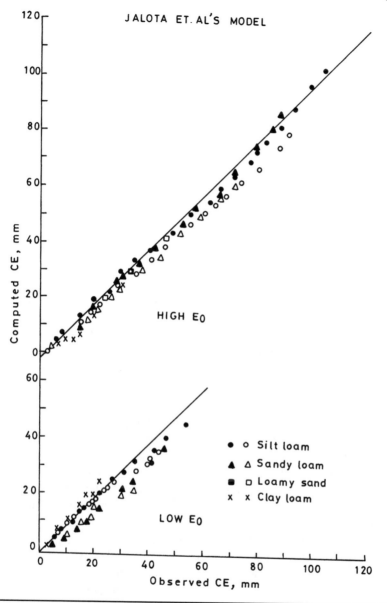

FIGURE 3.7. Continued

TABLE 3.7   Coefficient K of the Relation CE = $Kt^{0.5} - b$ as Influenced by Different Management Practices, E_0, and Soil Type

Treatments	Mulch rate (Mg ha⁻¹)					
	$E_0 = 10$ mm day⁻¹			$E_0 = 2.5$ mm day⁻¹		
	0	3	6	0	3	6
	Silt loam					
No tillage	13.5	15.8	13.5	8.8	9.4	6.5
Tillage	—	8.9	8.3	—	3.6	4.2
Undercutting	—	9.5	6.1	—	4.6	3.9
Straw, mixed	—	5.6	5.2	—	4.0	3.5
	Sandy loam					
No tillage	12.6	13.1	10.5	5.4	6.8	4.0
Tillage	—	12.1	8.2	—	8.1	3.2
Undercutting	—	10.1	10.6	—	9.1	3.3
Straw, mixed	—	7.7	5.2	—	5.8	3.1

Source: Prihar, Jalota, and Steiner, 1996.

soils under low E_0. But for silt loam and sandy loam soils under high E_0, the computed CE value exceeded the observed when the soil-limited stage was entered. CE one day after this stage was entered was overestimated, which continued until the end of the study, giving a constant difference. This is because the CE versus $(t - t_1)^{0.5}$ regressions gave appreciable negative intercepts (15-18 mm) and the value of α exceeded E_0. The latter is conceptually untenable because E_0 is the upper limit to the day 1 evaporation of the falling-rate stage. Alternatively, a continuous function relating CE with $t^{0.5}$ from the very beginning of evaporation after wetting—the modified square root of time relation (Jalota, Prihar, and Gill, 1988) explained previously—was tried. Use of the modified square root of time relation improved the agreement between the experimental values and those computed from regression coefficient K of Table 3.7 for all cases (Figure 3.7). This suggests that CE can be estimated satisfactorily with Ritchie's model for residue-covered, coarse-textured soils for all evaporativities and for all soils for low E_0. But for soils with higher water retention and transmission under high E_0, the model of Jalota, Prihar, and Gill (1988) was more appropriate. For straw-mixed and undercut treatments, CE values computed by both the models agreed closely with the observed values, regardless of soil and E_0.

Numerical Approach

Including surface residue as an intimate part of the soil-atmosphere continuum, Bristow et al. (1986) and van Doren and Allmaras (1978) suggested models to predict evaporation from residue-covered soil using daily weather data (short-wave radiation, maximum and minimum temperature), site (latitude), residue (area index, angle of distribution, reflectivity, and tranmissivity), and soil (bulk density and air entry potential) parameters. Flerchinger and Saxton (1989) proposed a physically based SHAW model, which integrates heat, water, and solute transfer through snow, residue, and soil represented as an individual node. For each node, flux equations were written in implicit finite-difference form and were solved iteratively. Though these models are difficult to evaluate, they are quite useful in improving our understanding of atmosphere-residue-soil interactions. Their application helps us to (1) understand processes within the system and (2) optimize management of residue and tillage to better utilize soil and water resources.

Summary

Evaporation loss from soil, a component of field water balance, is difficult to measure directly. It is often computed as the difference in the (measurable) gains and losses of water by soil during a specified time interval. More precise estimates are obtained by using various-sized lysimeters, which permit continuous or periodic weighments. The difference in the mass of the lysimeter before and after allowing evaporation represents evaporation loss because runoff and deep drainage are not permitted. Alternatively, scientists have attempted to estimate rates and amounts of evaporation from soil through models that range from purely empirical statistical relations to analytical and numerical solutions of the Richard's flow equation subject to (assumed) initial and boundary conditions. Available models concerning evaporation from untreated bare, tilled bare, and residue-covered soil have been reviewed in this chapter. Models based on analytical solutions of the flow equation largely suffer from unrealistic assumptions such as homogeneous soil profile, uniform initial soil wetness, and infinitely high evaporativity (E_0), leading to instantaneous drying at the surface. These analyses have shown that the finer the soil texture and the lower the E_0, the longer the duration of the constant-rate stage. Cumulative evaporation during the falling-rate stage has been reported to increase linearly with the square root of time. However, the coefficient of regression varied with soil water diffusivity and E_0. Because the constant-rate stage of evaporation in tilled soil is short-lived, the square root of time gave a strong linear relation with post-tillage cumulative evaporation. The regression coefficient depended upon depth of tillage and E_0. Efforts to simulate tillage with a layer of water-proofed aggregates assuming instant drying of the tilled layer are apparently unrealistic. The tilled layer starts drying from the top and does not dry to its full depth until several days after tillage. A considerable amount of liquid water moves through this layer during this period.

Numerical analyses have been used to model evaporation from soil using finite-difference or finite-element techniques involving soil water diffusivity, water characteristics of the soil, and initial distribution of water in the profile. Evaporation is computed as the integral of water loss from all layers during a given time interval, assuming or ensuring zero flux at the bottom boundary.

Evaporation from residue-covered soil has been modeled analytically as well as numerically. In the numerical procedure daily weather data (short-wave radiation, maximum and minimum temperature), site (latitude), residue (area index, angle of distribution, reflectivity, and tranmissivity), and soil (bulk density and air entry potential) parameters are needed. In the analytical procedure the reduction in the constant-rate stage evaporation rate caused by residue is first worked out from empirical relations. The soil is then considered to dry as bare soil at the reduced rate. Because of the specificity of the results of the analytical exercise to soil transmission characteristics, initial conditions, and E_0, it is difficult to arrive at generalized solutions. It appears that quasi-analytical and numerical approaches will be more rewarding in modeling evaporation from soil. If the current progress is maintained, it should soon be possible to predict soil evaporation with reasonable accuracy.

Solved Examples

Example 1

An infinitely deep soil column wetted to a water content (θ_i) of 0.24 m^3 m^{-3} is allowed to dry under infinite evaporativity. If the air water content of soil (θ_a) is 0.03 m^3 m^{-3} and the mean weighted diffusivity (D) is 17.3 cm^2 day^{-1}, compute the total evaporation and the evaporation rate at different time periods until 10 days after the commencement of evaporation.

Solution

Total evaporation (E) under the given conditions is approximated by the equation $E = 2(\theta_i - \theta_a)(Dt/\pi)^{0.5}$. Plugging in the relevant values of θ_i, θ_a, D, and time, we get total E

at day 1 = 2(0.24 − 0.03)(17.3 × 1/3.1416)$^{0.5}$ = 0.99 cm
at day 2 = 2(0.24 − 0.03)(17.3 × 2/3.1416)$^{0.5}$ = 1.39 cm
at day 3 = 2(0.24 − 0.03)(17.3 × 3/3.1416)$^{0.5}$ = 1.70 cm
at day 4 = 2(0.24 − 0.03)(17.3 × 4/3.1416)$^{0.5}$ = 1.97 cm

at day 5 = 2(0.24 − 0.03)(17.3 × 5/3.1416)$^{0.5}$ = 2.20 cm
at day 6 = 2(0.24 − 0.03)(17.3 × 6/3.1416)$^{0.5}$ = 2.41 cm
at day 7 = 2(0.24 − 0.03)(17.3 × 7/3.1416)$^{0.5}$ = 2.61 cm
at day 8 = 2(0.24 − 0.03)(17.3 × 8/3.1416)$^{0.5}$ = 2.79 cm
at day 9 = 2(0.24 − 0.03)(17.3 × 9/3.1416)$^{0.5}$ = 2.96 cm
at day 10 = 2(0.24 − 0.03)(17.3 × 10/3.1416)$^{0.5}$ = 3.12 cm

Likewise, the evaporation rate is obtained by plugging in the values in the equation ER = $(\theta_i - \theta_a)(D/\pi t)^{0.5}$

at day 1 = (0.24 − 0.03)(17.3/3.1416 × 1)$^{0.5}$ = 0.49 cm day^{-1}
at day 2 = (0.24 − 0.03)(17.3/3.1416 × 2)$^{0.5}$ = 0.39 cm day^{-1}
at day 3 = (0.24 − 0.03)(17.3/3.1416 × 3)$^{0.5}$ = 0.28 cm day^{-1}
at day 4 = (0.24 − 0.03)(17.3/3.1416 × 4)$^{0.5}$ = 0.25 cm day^{-1}
at day 5 = (0.24 − 0.03)(17.3/3.1416 × 5)$^{0.5}$ = 0.22 cm day^{-1}
at day 6 = (0.24 − 0.03)(17.3/3.1416 × 6)$^{0.5}$ = 0.20 cm day^{-1}
at day 7 = (0.24 − 0.03)(17.3/3.1416 × 7)$^{0.5}$ = 0.19 cm day^{-1}
at day 8 = (0.24 − 0.03)(17.3/3.1416 × 8)$^{0.5}$ = 0.17 cm day^{-1}
at day 9 = (0.24 − 0.03)(17.3/3.1416 × 9)$^{0.5}$ = 0.16 cm day^{-1}
at day 10 = (0.24 − 0.03)(17.3/3.1416 × 10)$^{0.5}$ = 0.16 cm day^{-1}

Example 2

A 95 cm long soil column wetted uniformly to initial wetness of 0.237 m^3 m^{-3} was allowed to dry under E_0 of 1.06 cm day^{-1}. Assuming a diffusivity function of the type $D = a \exp(b\theta)$ where $a = 2.70$ cm^2 day^{-1} and $b = 17.54$ cm^3 cm^{-3}, calculate the evaporation loss from the column for 10 days.

Solution

According to Gardner and Hillel's (1962) method, the evaporation rate is given by

ER = $-dW/dt = D(\theta)W\pi^2/4L^2$

where ER is evaporation rate (cm day^{-1}); W is total profile water storage (cm); t is time (days); D is diffusivity (cm^2 day^{-1}), a function of mean wetness, θ (cm^3 cm^{-3}); and L is the length of the wetted profile (cm). The mean wetness (θ) is related to total profile water storage (W) by $\theta = W/L$ or $W = \theta L$. Substituting appropriate values in the equation, $D = 2.70 \exp(17.54 × 0.237) = 172.47$ cm^2 day^{-1} and $\theta = 0.237$ cm^3 cm^{-3}, we get

1st day ER = $(172.47 \times 0.237 \times 95 \times 3.1416 \times 3.1416)/(4 \times 95 \times 95)$

$= 1.06$ cm day^{-1}

If the 1st day evaporation rate is more than the potential evaporation rate, we take the evaporation rate to be equal to evaporativity because on the 1st day it is controlled by evaporativity. At the end of day 1, $W = 22.5 - 1.06 = 21.44$ cm; $\theta = 21.44/95 = 0.2256$ cm^3 cm^{-3}; and $D = 2.70$ exp(17.54 $\times$ 0.2256) = 141.2 cm^2 day^{-1}. Therefore,

2nd day ER = $(141.2 \times 0.2256 \times 95 \times 3.1416 \times 3.1416)/(4 \times 95 \times 95)$

$= 0.830$ cm day^{-1}

At the end of day 2, $W = 21.44 - 0.83 = 20.61$ cm; $\theta = 20.61/95 = 0.2169$ cm^3 cm^{-3}; and $D = 2.70$ exp(17.54 $\times$ 0.2169) = 121.62 cm^2 day^{-1}. Therefore,

3rd day ER = $(121.62 \times 0.2169 \times 95 \times 3.1416 \times 3.1416)/(4 \times 95 \times 95)$

$= 0.685$ cm day^{-1}

At the end of day 3, $W = 20.61 - 0.685 = 19.925$ cm; $\theta = 19.925/95 = 0.2097$ cm^3 cm^{-3}; and $D = 2.70$ exp(17.54 $\times$ 0.2097) = 107.19 cm^2 day^{-1}. Therefore,

4th day ER = $(107.19 \times 0.2097 \times 95 \times 3.1416 \times 3.1416)/(4 \times 95 \times 95)$

$= 0.583$ cm day^{-1}

At the end of day 4, $W = 19.925 - 0.583 = 19.342$ cm; $\theta = 19.342/95 = 0.2036$ cm^3 cm^{-3}; and $D = 2.70$ exp(17.54 $\times$ 0.2036) = 96.3 cm^2 day^{-1}. Therefore,

5th day ER = $(96.3 \times 0.2036 \times 95 \times 3.1416 \times 3.1416)/(4 \times 95 \times 95)$

$= 0.509$ cm day^{-1}

At the end of day 5, $W = 19.342 - 0.509 = 18.833$ cm; $\theta = 18.833/95 = 0.1982$ cm^3 cm^{-3}; and $D = 2.70$ exp(17.54 $\times$ 0.1982) = 87.32 cm^2 day^{-1}. Therefore,

6th day ER = $(87.32 \times 0.1982 \times 95 \times 3.1416 \times 3.1416)/(4 \times 95 \times 95)$

$= 0.450$ cm day^{-1}

At the end of day 6, $W = 18.833 - 0.450 = 18.383$ cm; $\theta = 18.383/95 = 0.1935$ cm^3 cm^{-3}; and $D = 2.70$ exp(17.54 $\times$ 0.1935) = 80.41 cm^2 day^{-1}. Therefore,

7th day ER = $(80.41 \times 0.1935 \times 95 \times 3.1416 \times 3.1416)/(4 \times 95 \times 95)$
$= 0.404$ cm day^{-1}

At the end of day 7, $W = 18.383 - 0.404 = 17.979$ cm; $\theta = 17.979/95$ $= 0.1892$ cm^3 cm^{-3}; and $D = 2.70$ exp$(17.54 \times 0.1892) = 74.80$ cm^2 day^{-1}. Therefore,

8th day ER = $(74.80 \times 0.1892 \times 95 \times 3.1416 \times 3.1416)/(4 \times 95 \times 95)$
$= 0.368$ cm day^{-1}

At the end of day 8, $W = 17.989 - 0.368 = 17.610$ cm; $\theta = 17.610/95$ $= 0.1854$ cm^3 cm^{-3}; and $D = 2.70$ exp$(17.54 \times 0.1854) = 69.77$ cm^2 day^{-1}. Therefore,

9th day ER = $(69.77 \times 0.1854 \times 95 \times 3.1416 \times 3.1416)/(4 \times 95 \times 95)$
$= 0.336$ cm day^{-1}

At the end of day 9, $W = 17.610 - 0.336 = 17.274$ cm; $\theta = 17.274/95$ $= 0.1818$ cm^3 cm^{-1}; and $D = 2.70$ exp$(17.54 \times 0.1818) = 65.49$ cm^2 day^{-1}. Therefore,

10th day ER = $(65.49 \times 0.1818 \times 95 \times 3.1416 \times 3.1416)/(4 \times 95 \times 95)$
$= 0.309$ cm day^{-1}

Total evaporation for 10 days is

$1.06 + 0.830 + 0.685 + 0.583 + 0.509 + 0.450 + 0.404 + 0.368 + 0.336 + 0.309$
$= 5.534$ cm

Example 3

A soil is wetted to 100 cm depth to an average volumetric water fraction of 0.30. If the soil water diffusivity is 100 cm^2 day^{-1} and the potential evaporation rate is 0.60 cm day^{-1}, find the actual evaporation rate from soil.

Solution

Here $W = \theta L = 0.30 \times 100 = 30$ cm. According to the Gardner and Hillel (1962) method,

ER = $DW\pi^2/4L^2 = (100 \times 30 \times 3.1416 \times 3.1416)/(4 \times 100 \times 100) = 0.740$ cm day^{-1}

In this case the soil can supply water to the surface at the rate of 0.740 cm day^{-1}, which is greater than the potential evaporation rate (0.6 cm day^{-1}). So we will take the actual evaporation from the soil to be equal to the potential evaporation.

Example 4

If a soil with a mean weighted diffusivity of 200 cm^2 day^{-1} and an average water content of 0.1 cm^3 cm^{-3} in a 90 cm deep column is drying at an E_0 of 1.0 cm day^{-1}, find the actual evaporation rate.

Solution

$$W = \theta L = 0.10 \times 90 = 9.0 \text{ cm}$$

$$ER = DW\pi^2/4L^2 = (200 \times 9.0 \times 3.1416 \times 3.1416)/(4 \times 90 \times 90) = 0.548 \text{ cm day}^{-1}$$

Example 5

Calculate the potential evapotranspiration for natural conditions from the following data. Wind speed is 2000 m hr^{-1} at 1 m height and 2500 m hr^{-1} at 2 m, air temperature is 20°C at 1 m, and vapor pressure is 10 mb at 1 m height and 9 mb at 2 m. The density of air (ρ_a) at 20°C is 1.2 × 10^{-3} g cm^{-3}, Van Karman's constant (K) is 0.41, the ratio of the molecular weight of water to that of air (ε) is 0.662, and the atmospheric pressure (P_a) is 1013 mb.

Solution

The potential evapotranspiration (PET) is given by the equation

$$PET = -\left(\frac{\rho_a \varepsilon}{P_a}\right) K' \left(\frac{e_2 - e_1}{z_2 - z_1}\right)$$

To use this we first compute the eddy transfer coefficient (K'):

$$K' = \frac{K^2 \left(U_2 - U_1\right)\left(Z_2 - Z_1\right)}{\ln\left(Z_2 / Z_1\right)^2}$$

$$= \frac{0.41 \times 0.41(2500 - 2000)(2 - 1)}{\ln(2/1)^2} = 175 \text{ m}^2 \text{ hr}^{-1}$$

$$= 175 \times 10^4 \text{ cm}^2 \text{ hr}^{-1}$$

Now insert the appropriate values in the equation for PET:

$$PET = -\left[\frac{(1.2\times10^{-3})\times0.662}{1013}\right](175\times10^4)\frac{9-10}{200-100}$$

$$= 1.29\times10^{-2} \text{ g cm}^2 \text{ hr}^{-1} = 1.29\times10^{-2} \text{ cm hr}^{-1}$$

Example 6

Instruments set up to measure air temperature and vapor pressure at two heights, net radiation, and soil heat flow recorded the following data: Temperature (T) at $1 \text{ m} = 21°C$ and vapor pressure $(e) = 11$ mb; at 2m, $T = 20°C$ and $e = 10$ mb; $R_n = 330$ ly day^{-1}; and $G = -30$ ly day^{-1}. Find the latent heat of evapotranspiration (LE_t) using Bowen's energy balance ratio method. The heat capacity of air $= 0.24$ cal g^{-1} °C^{-1}; atmospheric pressure $(P_a) = 1013$ mb; and the ratio of the molecular weight of water to that of air $(\varepsilon) = 0.662$.

Solution

According to Bowen's energy balance ratio method,

$$LE_t = (Rn+G)/\left[1+\frac{C_p P_a(T_2-T_1)}{L\varepsilon(e_2-e_1)}\right]$$

Plugging in the values from recorded data, we get

$$LE_t = [330+(-30)]/\left[1+\frac{0.24\times1013\times(20-21)}{585\times0.622\times(10-11)}\right] = 180 \text{ ly day}^{-1}$$

Example 7

Estimate evaporation from bare untreated sandy loam soil for 10 days under an evaporativity of 10 mm day^{-1} with Ritchie's model $[CE_1 = U, CE_2 = \alpha(t-t_1)^{0.5} + U]$ with $U = 24$ mm and $\alpha = 13.44$ mm day$^{-0.5}$.

Solution

A U of 24 mm shows that evaporation occurred at a potential rate of 10 mm day^{-1} for 2.4 days. Hence, evaporation

after 1 day $= 10.0$ mm
after 2 days $= 20.0$ mm

after 3 days = $13.44(3 - 2.4)^{0.5} + 24.0 = 34.4$ mm
after 4 days = $13.44(4 - 2.4)^{0.5} + 24.0 = 41.0$ mm
after 5 days = $13.44(5 - 2.4)^{0.5} + 24.0 = 45.6$ mm
after 6 days = $13.44(6 - 2.4)^{0.5} + 24.0 = 49.5$ mm
after 7 days = $13.44(7 - 2.4)^{0.5} + 24.0 = 52.8$ mm
after 8 days = $13.44(8 - 2.4)^{0.5} + 24.0 = 55.8$ mm
after 9 days = $13.44(9 - 2.4)^{0.5} + 24.0 = 58.5$ mm
after 10 days = $13.44(10 - 2.4)^{0.5} + 24.0 = 61.1$ mm

Example 8

Calculate evaporation for 10 days from sandy loam soil covered with 3 Mg ha^{-1} wheat straw under an evaporativity of 10 mm day^{-1} with Ritchie's model if α for the soil is 11.9 mm day$^{-0.5}$. The relation for the ratio of potential evaporation rate from mulch–covered soil (E_m) to evaporativity (E_0) and mulch rate (MR) is $E_m/E_0 = 1.04 \exp[-(0.225MR + 0.0074E_0)]$, and for the duration of the constant-rate stage, $t_1 = 13.53E_0^{-1.037}$.

Solution

$$E_m = 10.0 \times 1.04 \exp\{-[(0.225 \times 3) + (0.0074 \times 10)]\} = 10.0 \times 1.04 \times 0.4728$$
$$= 4.922 \text{ mm day}^{-1}$$

$$t_1 = 13.53 \times 4.922^{-1.037} = 2.6 \text{ days}$$

$$U = E_m t_1 = 4.922 \times 2.6 = 12.8 \text{ mm}$$

Therefore, the evaporation obtained by plugging the appropriate values into the equation is

after 1 day = 4.9 mm
after 2 days = 9.8 mm
after 3 days = $11.9(3 - 2.6)^{0.5} + 12.8 = 17.6$ mm
after 4 days = $11.9(4 - 2.6)^{0.5} + 12.8 = 26.9$ mm
after 5 days = $11.9(5 - 2.6)^{0.5} + 12.8 = 31.2$ mm
after 6 days = $11.9(6 - 2.6)^{0.5} + 12.8 = 34.7$ mm
after 7 days = $11.9(7 - 2.6)^{0.5} + 12.8 = 37.8$ mm
after 8 days = $11.9(8 - 2.6)^{0.5} + 12.8 = 40.5$ mm
after 9 days = $11.9(9 - 2.6)^{0.5} + 12.8 = 42.9$ mm
after 10 days = $11.9(10 - 2.6)^{0.5} + 12.8 = 45.2$ mm

Note that the evaporation rate on the third day in Example 7 and on the third and fourth days in Example 8 (onset of falling-rate stage) exceeded the evaporativity.

Example 9

Using the method of Jalota, Prihar, and Gill (1988), calculate the duration of the constant-rate stage (t_f) and cumulative evaporation for 10 days from soil columns wetted to field capacity that are drying under an evaporativity of 10.0 mm day^{-1}. The mean weighted diffusivity (D) of the soil is 1707 mm^2 day^{-1}. K is related to D and E_0 as

$$K = 2.908 + 0.00298D + 1.032E_0 + 0.0006E_0D - 0.095E_0^2$$

Solution

This example is solved with the Jalota, Prihar, and Gill (1988) model:

$$
\begin{aligned}
CE_1 &= E_0t_f & \text{for } t < t_f \\
CE &= Kt^{0.5} - b & \text{for } t > t_f \\
t_f &= K^2/4E_0^2 \\
b &= K^2/4E_0
\end{aligned}
$$

We first work out the value of K by plugging the values of D and E_0 into the given relation. Thus,

$$
\begin{aligned}
K &= 2.908 + (0.00298 \times 1707) + (1.032 \times 10) + (0.0006 \times 1707 \times 10) \\
&\quad - (0.0095 \times 10 \times 10) = 19.056 \text{ mm day}^{-0.5}
\end{aligned}
$$

The duration of the constant-rate stage is

$$t_f = K^2/4E_0^2 = (19.056 \times 19.056)/(4 \times 10 \times 10) = 0.9 \text{ days}$$

$$b = (19.056 \times 19.056)/(4 \times 10) = 9.1 \text{ mm}$$

The evaporation computed by the equation $CE = Kt^{0.5} - b$ for different periods is

$$
\begin{aligned}
\text{after 1 day} &= 19.06(1)^{0.5} - 9.1 = 10.0 \text{ mm} \\
\text{after 2 days} &= 19.06(2)^{0.5} - 9.1 = 17.9 \text{ mm} \\
\text{after 3 days} &= 19.06(3)^{0.5} - 9.1 = 23.9 \text{ mm}
\end{aligned}
$$

after 4 days = $19.06(4)^{0.5} - 9.1 = 29.0$ mm
after 5 days = $19.06(5)^{0.5} - 9.1 = 33.5$ mm
after 6 days = $19.06(6)^{0.5} - 9.1 = 37.6$ mm
after 7 days = $19.06(7)^{0.5} - 9.1 = 41.3$ mm
after 8 days = $19.06(8)^{0.5} - 9.1 = 44.8$ mm
after 9 days = $19.06(9)^{0.5} - 9.1 = 48.0$ mm
after 10 days = $19.06(10)^{0.5} - 9.1 = 51.2$ mm

Example 10

Calculate evaporation from sandy loam soil tilled to 5 cm depth under an evaporativity of 10 mm day^{-1} using the Jalota, Prihar, and Gill (1988) method (see Example 9). The K for tilled soil is related to the thickness (S) of the tilled layer and E_0 as follows:

$$K = 10.82 + 0.38E_0 - 4.61S + 0.57S^2 + 0.000064SE_0$$

Solution

We first compute K for 5 cm deep cultivation and an E_0 of 10 mm day^{-1}:

$$K = 10.82 + (0.38 \times 10) - (4.61 \times 5) + (0.57 \times 5 \times 5) + (0.000064 \times 5 \times 10)$$
$$= 5.82 \text{ mm day}^{-0.5}$$

$t_f = (5.82 \times 5.82)/(4 \times 10 \times 10) = 0.08$ day

$b = (5.82 \times 5.82)/(4 \times 10) = 0.85$ mm

Plugging the values of K and b into the equation, we get evaporation

after 1 day = $5.82(1)^{0.5} - 0.85 = 5.0$ mm
after 2 days = $5.82(2)^{0.5} - 0.85 = 7.4$ mm
after 3 days = $5.82(3)^{0.5} - 0.85 = 9.2$ mm
after 4 days = $5.82(4)^{0.5} - 0.85 = 10.8$ mm
after 5 days = $5.82(5)^{0.5} - 0.85 = 12.2$ mm
after 6 days = $5.82(6)^{0.5} - 0.85 = 13.4$ mm
after 7 days = $5.82(7)^{0.5} - 0.85 = 14.5$ mm
after 8 days = $5.82(8)^{0.5} - 0.85 = 15.6$ mm
after 9 days = $5.82(9)^{0.5} - 0.85 = 16.6$ mm
after 10 days = $5.82(10)^{0.5} - 0.85 = 17.6$ mm

References

Acharya, C. L., and S. S. Prihar. 1969. Vapor losses through soil mulch at different wind velocities. *Agronomy Journal* 61:666–668.

Black, T. A., W. R. Gardner, and G. W. Thurtell. 1969. The prediction of evaporation, drainage, and soil water storage for a bare soil. *Soil Science Society of America Proceedings* 33:655–660.

Boast, C. W., and T. M. Robertson. 1982. A "micro-lysimeter" method for determining evaporation from bare soil: Description and laboratory evaluation. *Soil Science Society of America Journal* 46:689–696.

Boesten, J. J. T. I., and L. Stroosnijder. 1986. Simple model for daily evaporation from fallow tilled soil under spring conditions in a temperate climate. *Netherlands Journal of Agricultural Science* 34:75–90.

Bowen, I. S. 1926. The ratio of heat losses by conduction and by evaporation from any water surface. *Physics Review* 27:779–787.

Bristow, K. L., G. S. Campbell, R. I. Papendick, and L. F. Elliott. 1986. Simulation of heat and moisture transfer through a surface residue-soil system. *Agricultural and Forest Meteorology* 36:193–214.

Budagoveskly, E. I. 1962. *Evaporation of Water by Soil (Heat Balance of Forest Field)*. Moscow: Izdania Akadmia Nauk.

Camillo, P. J., R. J. Gurney, and T. J. Schmugge. 1983. A soil and atmosphere boundary layer model for evapotranspiration and soil moisture studies. *Water Resources Research* 19:371–380.

Covey, W. 1963. Mathematical study of first stage of drying of moist soil. *Soil Science Society of America Proceedings* 27:130–134.

Crank, J. 1956. *The Mathematics of Diffusion*. London: Oxford University Press.

Crank, J., and P. Nicholson. 1947. A practical method for numerical evaluation of solution of partial differential equations of the heat conduction type. *Proceedings of the Cambridge Philosophical Society of Mathematical and Physical Sciences* 43:50–67.

Flerchinger, G. N., and K. E. Saxton. 1989. Simultaneous heat and water model of a freezing snow-residue-soil system. *Transactions of the American Society of Agricultural Engineers* 32:565–571.

Fritton, D. D., D. Kirkham, and R. H. Shaw. 1967. Soil water and chloride redistribution under various evaporation potentials. *Soil Science Society of America Proceedings* 31:599–603.

Fuchs, S. M., and C. B. Tanner. 1967. Evaporation from a drying soil. *Journal of Applied Meteorology* 6:852–857.

Gardner, H. R. 1974. Prediction of water loss from a fallow field soil based on soil water flow theory. *Soil Science Society of America Proceedings* 38:379–382.

Gardner, W. R. 1959. Solutions of flow equation for the drying of soils and other porous media. *Soil Science Society of America Proceedings* 23:183–187.

Gardner, W. R., and M. Fireman. 1958. Laboratory studies of evaporation from soil columns in the presence of a water table. *Soil Science* 85:244–249.

Gardner, W. R., and D. I. Hillel. 1962. The relation of external evaporation conditions to the drying of soils. *Journal of Geophysical Research* 67:4319–4325.

Gardner, W. R., and M. S. Mayhugh. 1958. Solutions and tests of the diffusion equation for the movement of water in soil. *Soil Science Society of America Proceedings* 22:197–201.

Gill, B. S. 1992. Interactive effects of paddy straw mulch rates and tillage depths on evaporation reduction in relation to soil type and atmospheric evaporativity. M.Sc. thesis, Punjab Agricultural University, Ludhiana, India.

Gill, K. S., and S. S. Prihar. 1983. Cultivation and evaporativity effects on drying patterns in sandy loam soil. *Soil Science* 135:366–377.

Hadas, A. 1975. Drying of layered soil columns under nonisothermal conditions. *Soil Science* 119:143–148.

Hall, A. E., and C. Dancette. 1978. Analysis of fallow-farming systems in semi-arid Africa using a model to simulate the hydrologic budget. *Agronomy Journal* 70:816–823.

Hammel, J. E., R. I. Papendick, and G. S. Campbell. 1981. Fallow tillage effects on evaporation and seed zone water content in a dry summer climate. *Soil Science Society of America Journal* 45:1016–1022.

Hanks, R. J. 1958. Vapor transfer in dry soil. *Soil Science Society of America Proceedings* 22:372–374.

Hanks, R. J., H. R. Gardner, and M. L. Fairbourn. 1967. Evaporation of water from soil as influenced by drying with wind or radiation. *Soil Science Society of America Proceedings* 31:593–598.

Hanks, R. J., and N. P. Woodruff. 1958. Influence of wind on vapor transfer through soil, gravel, and straw mulches. *Soil Science* 86:160–164.

Heller, J. P. 1968. The drying through the top surface of a vertical porous column. *Soil Science Society of America Proceedings* 32:778–785.

Hillel, D., and A. Hadas. 1972. Isothermal drying of structurally layered soil columns. *Soil Science* 113:495–498.

Hillel, D., and H. Talpaz. 1977. Simulation of soil water dynamics in layered soils. *Soil Science* 123:54–62.

Hillel, D., C. H. M. van Bavel, and H. Talpaz. 1975. Dynamic simulation of water storage in fallow soil as affected by mulch of hydrophobic aggregates. *Soil Science Society of America Proceedings* 39:826–833.

Idso, S. B., and R. D. Jackson. 1969. Thermal radiations from the atmosphere. *Journal of Geophysical Research* 74:5397–5403.

Idso, S. B., R. J. Reginato, and R. D. Jackson. 1979. Calculation of evaporation during the three stages of drying. *Water Resources Research* 15:487–488.

Jackson, R. D., B. A. Kimball, R. J. Reginato, and F. S. Nakayama. 1974. Diurnal soil water evaporation: Comparison of measured and calculated water fluxes. *Soil Science Society of America Proceedings* 38:861–866.

Jackson, R. D., S. B. Idso, and R. J. Reginato. 1976. Calculation of evaporation rate during the transition from energy limiting to soil limiting phases using albedo data. *Water Resources Research* 12:23–26.

Jalota, S. K. 1984. Drying pattern of bare and tilled silt loam, sandy loam, and loamy sand soils as affected by zero time water profiles and evaporativity. Ph.D. diss., Punjab Agricutural University, Ludhiana, India.

_____. 1990. Suitability of techniques for predicting falling rate stage evaporation in relation to soil type and evaporativity. *Journal of the Indian Society of Soil Science* 38:380–384.

_____. 1994. Evaporation parameter in relation to soil texture and atmospheric evaporativity to predict evaporation from bare soil. *Journal of the Indian Society of Soil Science* 42:178–181.

_____. 1995. Estimating soil evaporation parameter in relation to tillage depth, soil type, and evaporativity. *Journal of the Indian Society of Soil Science* 43:667–668.

Jalota, S. K., and S. S. Prihar. 1986. Effect of atmospheric evaporativity, soil type, and redistribution time on evaporation from bare soil. *Australian Journal of Soil Research* 24:357–366.

_____. 1987. Observed and predicted evaporation trends from a sandy loam soil under constant and staggered evaporativity. *Australian Journal of Soil Research* 25:243–249.

Jalota, S. K., S. S. Prihar, and K. S. Gill. 1988. Modified square root of time relation to predict evaporation trends from bare soil. *Australian Journal of Soil Research* 26:281–289.

Kijne, J. W. 1973. Evaporation from bare soils. In *Physical Aspects of Soil Water and Salts in Ecosystems*, edited by D. Hadas, D. Swartzendruber, P. E. Rijtema, M. Fuchs, and B. Yaron, pp. 221–226. London: Chapman & Hall.

Klute, A., F. D. Whisler, and E. J. Scott. 1965. Numerical solution of non-linear diffusion equation for water flow in a horizontal column of finite length. *Soil Science Society of America Proceedings* 29:353–356.

Krischer, O., and H. Rohnalter. 1940. *Warmleitang und Damp of Diffusion in feuchten Gutern*. VDI Forschungsheet 402.

Lascano, R. J., and C. H. M. van Bavel. 1983. Experimental verification of a model to predict soil moisture and temperature profiles. *Soil Science Society of America Journal* 47:441–448.

_____. 1986. Simulation and measurement of evaporation from bare soil. *Soil Science Society of America Journal* 50:1127–1132.

Linden, D. R. 1982. Predicting tillage effects on evaporation. In *Predicting Tillage Effects on Soil Physical Properties and Processes*, pp. 117–132. ASA Special Publication no. 44. Madison, Wis.: American Society of Agronomy, Soil Science Society of America.

McIlroy, I. C., and D. E. Angus. 1964. Grass, water, and soil evaporation at Aspendale. *Agricultural Meteorology* 1:201–204.

Monteith, J. L. 1965. Evaporation and environment. In *The State and Movement of Water in Living Organisms*, pp. 205–234. Symposia of the Society for Experimental Biology, vol. 19. New York: Academic Press.

Penman, H. L. 1948. Natural evaporation from open water, bare soil, and grass. *Proceedings of the Royal Society London* 93:120–195.

_____. 1956. An introductory survey. *Netherlands Journal of Agricultural Science* 4:9–29.

Philip, J. R. 1957. Evaporation and moisture and heat fields in the soil. *Journal of Meteorology* 13:354–366.

Philip, J. R., and D. A. de Vries. 1957. Moisture movement in porous materials under temperature gradients. *Transactions of the American Geophysical Union* 38:222–232.

Priestley, C. H. B., and R. J. Taylor. 1972. On the assessment of surface heat flux and evaporation using large-scale parameters. *Monthly Weather Review* 100:81–92.

Prihar, S. S., S. K. Jalota, and J. L. Steiner. 1996. Residue management for evaporation reduction in relation to soil type and evaporativity. *Soil Use and Management* 12:150–157.

Prihar, S. S., and D. M. van Doren Jr. 1967. Mode of response of weed-free corn to post-planting cultivation. *Agronomy Journal* 59:513–516.

Pruit, W. O., and D. E. Angus. 1960. Large weighing lysimeter for measuring evapotranspiration. *Transactions of the American Society of Agricultural Engineers* 3:13–18.

Reynolds, W. D., and G. K. Walker. 1984. Development and validation of a numerical model simulating evaporation from short cores. *Soil Science of America Journal* 48:960–969.

Ritchie, J. T. 1972. Model for predicting evaporation from a row crop with incomplete cover. *Water Resources Research* 8:1204–1213.

Roesma, A. 1975. *A Mathematical Model for Simulation of the Thermal Behaviour of Bare Soil Based on Heat and Moisture Transfer.* NIWARS Publication no. 11. Delft, Netherlands: NIWARS.

Rose, C. W. 1968. Evaporation from bare soil under high radiation conditions. In *Transactions of the 9th International Congress of Soil Science*, pp. 57–68. Adelaide, Australia: International Society of Soil Science.

_____. 1969. Water transport in soil by evaporation and infiltration. In *Water in Unsaturated Zone: Proceedings of the Wageningen Symposium*, edited by P. E. Rijtema and H. Wassink, pp. 171–181. Wageningen, the Netherlands.

Rowse, H. R. 1975. Simulation of water balance of soil columns and fallow soils. *Journal of Soil Science* 26:337–349.

Singh, R., M. C. Oswal, and Jagan Nath. 1985. Prediction of falling rate evaporation from bare soils. *Journal of the Indian Society of Soil Science* 33:1–4.

Stroosnijder, L., and D. Kone. 1982. Le bilan d'eau du sol. In *La productivité des paturages saheliens*, edited by F. W. T. Penning de Vries and M. A. Djiteye, pp. 133–165. Agricultural Research Report 918. Wageningen, the Netherlands.

Thornthwaite, C. W., and B. Holzman. 1942. *Measurement of Evaporation from Land and Water Surfaces.* U.S. Department of Agriculture, Technical Bulletin no. 817. Washington, D.C.: Government Printing Office.

Tselishcheva, L. K. 1965. Influence of thickness and free porosity of the dried layer on evaporation. *Soviet Soil Science*, pp. 257–260.

van Bavel, C. H. M. 1966. Potential evaporation: The combination concept and its experimental verification. *Water Resources Research* 2:455–467.

van Bavel, C. H. M., and D. I. Hillel. 1976. Calculating potential and actual evaporation from a bare soil surface by simulation of concurrent flow of water and heat. *Agricultural Meteorology* 17:453–476.

van Bavel, C. H. M., and L. E. Myers. 1962. An automatic weighing lysimeter. *Agricultural Engineering* 43:580–583.

van Doren, D. M., Jr., and R. R. Allmaras. 1978. Effect of residue management practices on the soil physical environment, microclimate, and plant growth. In *Crop Residue Management Systems*, pp. 49–83. Agronomy Monograph no. 3. Madison, Wis.: American Society of Agronomy.

van Keulen, H., and D. Hillel. 1974. A simulation study of the drying front phenomenon. *Soil Science* 118:270–273.

van Keulen, H., and C. G. E. M. van Beek. 1971. Water movement in layered soils: A simulation model. *Netherlands Journal of Agricultural Science* 19:138–153.

von Rosenberg, D. U. 1969. *Methods for the Numerical Solution of Partial Equations*. New York: Elsevier.

Willis, W. O., and J. J. Bond. 1971. Soil water evaporation: Reduction by simulated tillage. *Soil Science Society of America Proceedings* 35:526–52.

4

Evaporation Reduction by Tillage

Shallow tillage is often advocated for reducing evaporation losses from soil, but its role in doing so is not unequivocally understood. The concept of tillage for moisture conservation originated from the famous laboratory experiments of King carried out in the United States in Wisconsin toward the end of the nineteenth century (see Baver, 1966). These experiments showed that soil mulch created by loosening a shallow layer on top of 120 cm tall columns that were kept in contact with free water at the bottom caused a large reduction in evaporation losses. In a pot culture study in Utah, 2.5 cm deep cultivation reduced evaporation losses by 34%, 13%, and 63% in sand, sanpete clay, and clay, respectively (Widstoe, 1909). Similar results had been obtained by Woolney and his school in Germany (Eser, 1884). Based on these and other similar results, dust mulch was advocated for conserving soil moisture against evaporation.

Later, however, the theory that dust mulch created by tillage reduced evaporation from soil was challenged by the results of the field experiments of Call and Sewell (1917) and Veihmeyer (1927) under weed-free conditions. It was argued that in the absence of a water table (a source of steady, upward supply), most of the water that could be lost by evaporation had already been lost by the time the soil mulch was created by tillage. Jacks, Brind, and Smith (1955) concluded from the then available literature on the subject that soil mulch did not reduce evaporation except by killing weeds. But there were reports in the literature that did show that soil mulch helped in reducing evaporation. For example, in a four-year field study, Singh and Nijhawan (1944) found evidence of moisture conservation by soil mulch created in situ. Benoit and Kirkham (1963) reported reduced evaporation with soil mulch applied extraneously to soil columns wetted to field capacity. However, the experimental setup of Benoit and Kirkham (1963) did not simulate the soil mulch created by in situ tillage. Prihar and van Doren (1967) determined evaporation losses from 15 cm diameter, 25 cm deep cans filled with silt loam soil and placed in tar-paper-lined holes between rows of growing corn (*Zea mays* L.) with their tops flush with the soil surface. Stirring the surface 5 cm layer with a garden hoe after each rain reduced short-term (3–5 day periods) evaporation losses by almost 50%. Aujla and Cheema (1983) reported that one shallow cultivation of a bare loamy sand field after cessation of the monsoon rains in September reduced evaporation loss by 20 mm over a period of 46 days. Minhas, Khosla, and Prihar (1986) determined the effect of 5 cm deep tillage on evaporation loss from a silt loam (field) soil employing the zero-flux-plane method of Ehlers and van Der Ploeg (1976). The plots were wetted sufficiently deep. Evaporation rates 18 days after tillage averaged 1.67 mm day^{-1} in the tilled and 2.16 mm day^{-1} in the untilled plots, and evaporativity (E_0) averaged 3.17 mm day^{-1}. Under an average E_0 of 8.47 mm day^{-1}, the tilled and untilled plots lost 3.80 and 4.26 mm water per day, respectively. Unger (1984) concluded that in areas where rainfall is frequent, clean tillage may reduce infiltration, and an additional tillage operation is required after each significant rainstorm. Thus, dust mulches are effective primarily in regions having distinct wet and dry seasons.

General Trends

Tillage generally increases the evaporation rate from soil during the initial period of drying and reduces it later, for reasons explained in Chapter 2. A number of laboratory (Prihar, Singh, and Sandhu, 1968; Willis and Bond, 1971; Gill et al., 1977; Gill and Prihar, 1983; Prihar and Jalota, 1988) and field (Papendick, Lindstrom, and Cochran, 1973; Jalota and Prihar, 1979; Aujla and Cheema, 1983; Chaudhary and Acharya, 1993) experiments have shown that the effect of tillage-induced soil mulch on evaporation reduction is governed by soil type, E_0, initial wetness, time of creation of soil mulch, and thickness and clod size distribution of the mulch. Moreover, the magnitude of the benefit is time variant. Therefore, soil water conservation by reducing evaporation can be achieved only if the effects of these variables are known.

Effect of Soil Type

In soils that tend to retain water and in which the hydraulic conductivity decreases gradually with decreasing water content, the surface soil remains moist for long periods because the high rate of upward flow continues to replenish the water lost from the upper layers. In such cases the upward flow can be retarded by loosening the top several centimeters of soil by tillage. This hastens the development of a dry surface layer and, thus, reduces evaporation from the tilled, compared with the untilled, soil.

On the other hand, in coarse-textured soils of low retentive capacity, a dry layer forms at the surface irrespective of tillage because the hydraulic conductivity decreases sharply with decreasing surface layer water content, and the upward flow is inadequate to replenish the water lost from the upper layer. In such soils the advantage of tillage in breaking up the capillary action is considerably smaller than in fine-textured soils. The limited number of (controlled) experiments involving more than one soil (Gill et al., 1977; Prihar, Singh, and Sandhu, 1968; Jalota and Prihar, 1990b; Jalota, 1995) have corroborated this line of reasoning. In a 30-day study, evaporation rates from (5 cm deep) tilled and untilled soil columns averaged 1.98 and 3.13 mm day^{-1} for silt loam and 1.41 and 2.13 mm day^{-1} for sandy loam soil, respectively, under an E_0 of 13.0 mm day^{-1} (Jalota and Prihar, unpublished). Obviously, the shallow tillage was more effective in water conservation for fine-textured soil than for coarse-textured soil.

Effect of Evaporativity

Under conditions of comparable soil type and initial wetness, high E_0 enhances the rate of water loss from soil surface layers. This reduces the water content and hydraulic conductivity of the upper soil layers, but at the same time it increases the hydraulic gradient. The latter tends to sustain the upward flow for some time. Eventually, however, the flow from lower layers fails to replenish the water loss from the surface, and surface layers become dry (Papendick, Lindstrom, and Cochran, 1973). The thickness of the dry layer that develops at the surface is just sufficient to balance the upward rate of flow. If a layer of higher resistance is created by accelerating drying, the evaporation rate will decrease. Thus, shallow tillage, which hastens the formation of a dry layer, results in substantial reduction in evaporation over short periods. But with continued drying, a dry layer also develops at the surface of untilled soil and the evaporation rate is reduced, often below that from the tilled soil. Therefore, under high E_0, evaporation reduction with tillage peaks after a certain length of time and declines thereafter. After a sufficiently long time the cumulative evaporation (CE) from tilled soil may equal or exceed that from untilled soil (Wiese and Army, 1958).

When the E_0 is low, the upward flow replenishes the small amount of water that is lost and keeps the soil surface moist for a long time. Penman (1941) showed that a low and steady demand on soil water supply could be met for a long time. Tillage reduced the upward flow by breaking the capillaries and, thus, reduced the evaporation rate compared with the untilled soil. Since a dry layer does not develop on untilled soil for a long time, evaporation reduction with tillage under low E_0 continues to increase for much longer periods compared with that under high E_0. Jalota and Prihar (1990b) observed that evaporation reduction with tillage of sandy loam soil peaked at 10 days after tillage under E_0 of 15.1 mm day^{-1} but continued to increase for 30 days under E_0 of 4.6 mm day^{-1}.

Effect of Time and Type of Tillage

Hanks and Gardner (1965) demonstrated that changes in diffusivity in the wet range have a large influence on the CE from soil. Since the $K(\theta)$ and $D(\theta)$ relations of a fine-textured soil differ a great deal from those of a coarse-textured soil, it is to be expected that the soil type and wetness at which $D(\theta)$ is changed with tillage will influence the course of evaporation. Willis and Bond (1971) tilled sandy loam soil columns exposed to an E_0 of 9.6 mm day^{-1} at 1, 4, 7, and 18 days after wetting and found that the earliest tillage gave the maximum reduction in evaporation. They also observed that when tillage was delayed for 18 days, 7.5 cm deep tillage tended to be more beneficial than 2.5 cm deep tillage. In later experiments Gill et al. (1977) observed that the effect of the time of tillage also varied with soil type, tilth, and E_0. They showed that under low E_0 tillage of sandy loam soil 5 and 11 days after wetting caused greater evaporation reduction than tillage at 18 and 42 days after wetting but that under high E_0 the earliest tillage was most effective. Interestingly, in the silty clay loam the second time of tillage (11 days after wetting) was the most effective (Table 4.1). It is postulated that when fine-textured soil is tilled in a wet condition under high E_0, it tends to reconsolidate and reestablish the capillarity. More recently this has been corroborated by Chaudhary and Acharya (1993) under field conditions. Further, coarse tilth with mean weight diameter (MWD) of clods ranging from 45 to 54 mm was found more effective in reducing evaporation loss when tillage was performed early, whereas fine tilth (MWD 7–13 mm) was more effective with delayed tillage. Thus, for a given time of tillage, the characteristics of the tilled layer, such as porosity, clod size distribution, and depth of tillage, influence the extent and time course of evaporation loss.

TABLE **4.1** Evaporation Loss for 67 Days Under E_0 of 2 mm Day^{-1} and for 52 Days Under E_0 of 9.8 mm Day^{-1} in Relation to Time of Tillage of Silty Clay Loam (Sicl) and Sandy Loam (Sl) Soils

Time of tillage, days after wetting	Evaporation (mm)			
	$E_0 = 2.0$ mm day^{-1}		$E_0 = 9.8$ mm day^{-1}	
	Sicl	Sl	Sicl	Sl
5	60.6a	56.8a	107.5c	85.1a
11	56.7a	58.1a	94.5a	86.4ab
18	60.2a	63.2b	102.8ab	88.4ab
42–43	69.3b	73.6c	104.2bc	88.5ab
Untilled	83.6c	77.6d	103.7abc	90.0b

Source: Gill et al., 1977.

Note: Any two means not followed by the same letters are significantly different at the 5% level.

Effect of the Interaction between Soil Type and Evaporativity

Soil type has been observed to interact strongly with evaporativity for moisture conservation by tillage (Prihar, Singh, and Sandhu, 1968; Gill et al., 1977; Jalota, 1984). Soil mulch created by tillage was found beneficial for moisture conservation on all soils under low E_0. But under high E_0 the benefit of tillage was confined to fine-textured soils. In fact, various combinations of E_0 and soil type yield a range of longevity and magnitude of evaporation reduction by tillage. The two extremes are represented by a fine-textured soil under low E_0 and a coarse soil under high E_0. When the soil is coarse and E_0 is high, a rapid decrease in surface water content is accompanied by a sharp decrease in hydraulic conductivity, and a dry layer is formed rapidly whether the soil is stirred or not. Under such conditions, tillage may hasten the formation of a dry layer only slightly and, hence, offers little benefit for moisture conservation (Prihar, Singh, and Sandhu, 1968). Interestingly, tillage of loamy sand was observed to have a negative effect on moisture conservation under high E_0 (Jalota, 1984; Jalota and Prihar, 1990a). However, where E_0 is low, the liquid flow, even in coarse-textured soils, continues for some time, and the hastening of the formation of a dry layer through tillage helps in reducing evaporation to some extent (Figure 4.1). On fine-textured soils under low E_0 cumulative reduction continued to increase for longer periods because the formation of a dry surface layer in the untilled soil took longer.

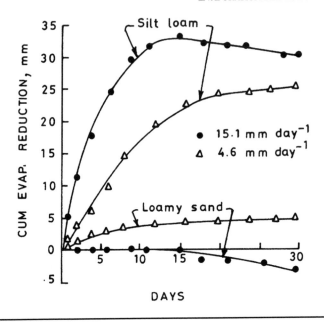

FIGURE **4.1.**   Evaporation reduction by tillage as a function of time in silt loam and loamy sand soil under two evaporativities (Jalota and Prihar, 1990a)

Effects of the Characteristics of the Tilled Layer

Air Porosity

The porosity of mulch has a direct relation to evaporation loss (Hanks, 1958; Acharya and Prihar, 1969). With an increase in the porosity fraction of the tilled layer from 0.50 to 0.64, evaporation increased by 25% (Allmaras, 1967). The reduction in porosity increases the diffusion resistance to vapor flow by making the flow path more tortuous. On this basis, compaction of the dry tilled layer is advocated to lower evaporation losses. However, another effect of reduced porosity is to increase the thermal conductivity of the tilled layer (Papendick, Lindstrom, and Cochran, 1973), which would tend to increase evaporation. On the other end of the scale, a decrease in thermal conductivity with an increase in porosity would enhance the vapor exchange by increasing air turbulence (Hanks and Woodruff, 1958; Acharya and Prihar, 1969). Therefore, to achieve maximum evaporation reduction, we must look for that threshold porosity of the tilled layer which reduces thermal conductivity without increasing the turbulent flow of air.

Since porosity of the tilled layer is a function of clod size distribution, a number of researchers have attempted to define the optimum clod size range

TABLE 4.2 Optimum Clod Sizes for Evaporation Reduction as Given by Different Workers

Reference	Clod size diameter (mm)
Wadham, 1944	<10.0
Ramacharlu, 1957	0.8
Holmes, Greacen, and Gurr, 1960	2.5
Scotter and Raats, 1969	0.3–0.6
Hillel and Hadas, 1972	0.5
Heinonen, 1985	0.5–2.0

Source: Jalota and Prihar, 1990a.

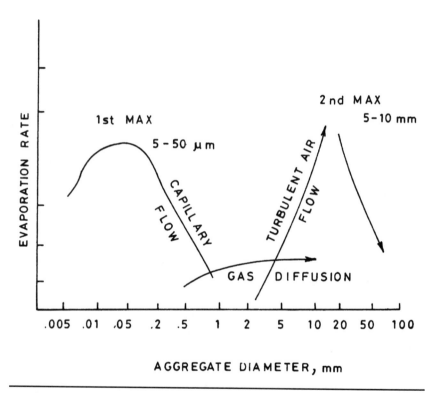

FIGURE 4.2. Schematic representation of evaporation rate and water transport process in relation to aggregate size (Heinonen, 1985)

for maximum evaporation reduction (Table 4.2). The inconsistency in the optimal size of aggregates is attributed to variable experimental conditions.

Heinonen (1985) showed schematically the main processes of water loss associated with the size of aggregates (Figure 4.2). The first peak of the evaporation rate occurred for aggregates of 0.05 mm diameter because of the high hydraulic conductivity of the aggregates. Increasing the aggregate size from 0.5 to 5.0 mm increased the air-filled porosity and slightly increased loss by gas diffusion. Wind-induced turbulence in soil pores started when the aggregate size was increased to 3–5 mm and became dominant for aggregates exceeding 10 mm diameter. The maximum drying rate was obtained in 5–10 mm diameter and larger clods. The surface area became a limiting factor for evaporation reduction, especially if hydraulic conductivity within the clod was low. In fact, the optimal clod size depends on the properties of the soil and wind gustiness.

Thickness of the Tilled Layer

Using surfactant-treated soil mulch Hanks (1958), Acharya and Prihar (1969), and Hillel, van Bavel, and Talpaz (1975) showed that evaporation loss was inversely related to mulch thickness. When the depth was less than a certain threshold value, evaporation was limited by external conditions rather than by mulch (Gardner and Fireman, 1958). Deeper tillage reduces evaporation to a greater extent by providing a stronger barrier to vapor diffusion (Willis and Bond, 1971; Gill and Prihar, 1983).

Actually, the tilled layer starts drying from the top and the drying front moves downward into the soil until the dry layer is sufficiently thick to reduce the loss rate to the point that it is balanced by unsaturated upward flow dictated by soil and climatic conditions. Jalota and Prihar (1990b) computed the thickness of an (absolutely) dry layer that permitted the same evaporation (vapor) loss as a tillage-induced soil mulch at a given point of time. The thickness of the dry layer was found to increase with time at least until 30 days after tillage under low E_0 of 4.9 ± 1.1 mm day^{-1} in silt loam and sandy loam soils. But under high E_0 of 15.3 ± 2.6 mm day^{-1}, it attained the maximum value of 0.9 cm after 19 days in sandy loam soil in coarse tilth (Figure 4.3). However, in fine tilth it increased up to 30 days. This shows that drying from the top down does not extend to the full depth of tillage. These results and those of Papendick, Lindstrom, and Cochran (1973) reveal that the formation of the dry layer is a dynamic process, and its rate of formation and effective depth are governed by E_0, the transmission properties of the soil, and the roughness of the tilled layer.

The lower (computed) thickness of dry soil mulch than that of the tilled layer indicated that, unlike in the case of completely dry soil, water in the tilled layer does not move as vapor alone. Possibly, the liquid flow from the undisturbed layer continued through or within the tilled layer, albeit at a lower rate than in the untilled soil. Evidently, therefore, a shallower tillage could be as effective in reducing evaporation as 5 cm deep tillage, provided the liquid

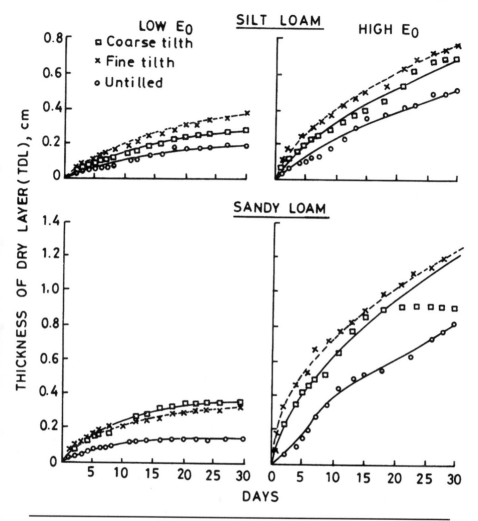

FIGURE 4.3. Thickness of (computed) dry layer as a function of time as affected by tilth, soil type, and E_0 (Jalota and Prihar, 1990b)

flow from lower layers to the tilled zone is completely cut off through accelerated drying of the tilled layer with additional stirring. This hypothesis was verified through experimentation, and the results are shown in Figure 4.4.

Compared with 5 cm deep tillage, an additional stirring of the 2 cm deep tilled layer after one day of tillage lowered the evaporation loss after 30 days by 7 and 2.5 mm in silt loam and sandy loam soils, respectively, under an E_0 of 4.9 mm day^{-1} and by 12 mm in silt loam under an E_0 of 15.3 mm day^{-1}. In sandy loam, 5 cm deep tillage was more effective than 2 cm under higher E_0.

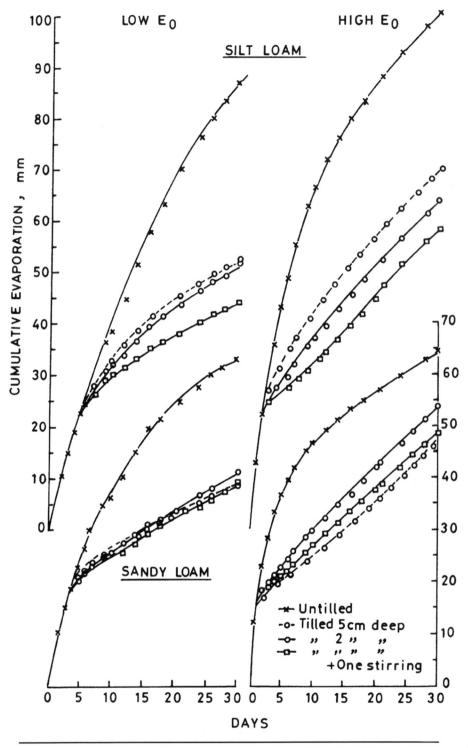

FIGURE 4.4. Time pattern of cumulative evaporation as affected by tillage manipulations, soil type, and E_0 (Jalota and Prihar, 1990b)

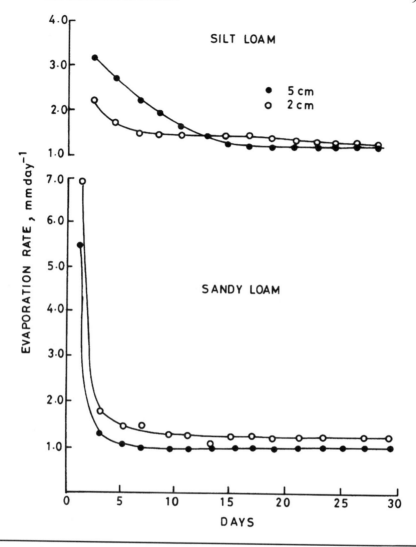

FIGURE **4.5.** Time pattern of evaporation rate under an E_0 of 15.3 mm day^{-1} as affected by depth of tillage (Jalota and Prihar, 1990b)

Interaction between Tillage Depth and Soil Type

The depth of tillage also exhibited an interaction with soil type (Jalota and Prihar, 1990b). Under high E_0, 2 cm deep tillage was found more effective than 5 cm deep tillage in the silt loam, while the reverse was true in the sandy loam (Figure 4.4). In the silt loam, shallower loosening caused more rapid drying of the surface layer and was initially more effective in reducing evaporation than the deeper loosening (Figure 4.5).

With the deeper (5 cm) tillage, the dry layer developed more slowly because the lower part of the tilled layer supplied water for evaporation. However, the dry layer thickened with time, and by the 12th day a thicker dry layer had developed with the deeper tillage than with the shallower tillage. Consequently, the evaporation rate with 5 cm deep tillage became lower than that with 2 cm deep tillage. In the sandy loam this phenomenon was absent. Apparently, the stirred layer dried rapidly, irrespective of the depth of tillage. On account of its lower water retentivity and lower diffusivity, the lower part of the tilled soil (in deeper tillage) did not transmit water fast enough to the upper part. Continuity of the upward liquid supply is also evident from results obtained by Jalota and Prihar (1992), who observed a higher rate of water movement through the junction of tilled and untilled soil than through a layer of waterproof dry aggregates of the thickness equal to that of the tilled layer (Figure 4.6).

The existence of this phenomenon is also supported by the continuity of the moisture profile in the tilled soil at 30 days of drying after tillage and the profile's distinct discontinuity when a layer of waterproof dry aggregates of the same thickness as tillage depth has been added (Figure 4.7). The water content of the lower portion of the tilled layer decreased only when the replenishment rate by unsaturated flow from moist soil below fell short of the loss rate dictated by external conditions (Papendick, Lindstrom, and Cochran, 1973).

The total evaporative flux from tilled soil comprises three components: vapor flux and liquid flux across the junction and loss from the tilled zone per se. Jalota and Prihar (1992) determined the magnitude and time trends of these fluxes in tilled columns of silt loam and sandy loam soils under evaporativities of 3.6 and 15.6 mm day^{-1}. During the initial period of drying after tillage, the major portion of the water loss in tilled soil came from the tilled zone per se (Figure 4.8). As drying proceeded, the contribution from the tilled layer declined and became negligible. With a decline in evaporative flux from the tilled zone, fluxes in liquid and vapor phases started from the untilled soil below the tilled zone. The loss of water entirely in vapor phase began when the tilled layer had dried out.

At the end of 30 days of drying, the liquid flux across the tilled layer, computed from the numerical solution of the flow equation and mass balance, in the tilled silt loam had contributed 72% to the total flux under low E_0 and 42% under high E_0. In the sandy loam it had contributed 20% under low and 31% under high E_0 (Table 4.3).

These results reveal that the previous assumptions of complete destruction of capillarity in tilled soil, instant drying of the tilled layer, and movement

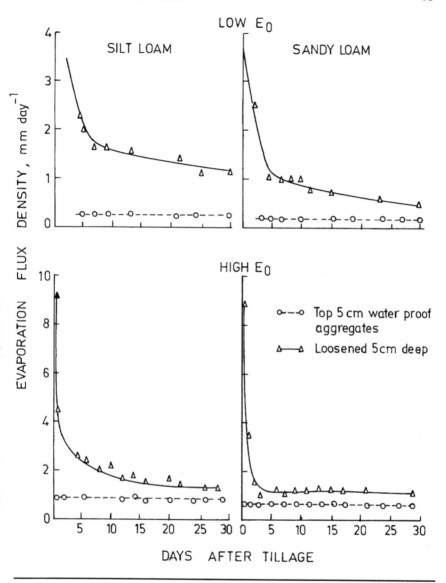

FIGURE 4.6. Time course of evaporative flux density from silt loam and sandy loam under low and high E_0 as affected by tillage and waterproof aggregates on the surface (Jalota and Prihar, 1992)

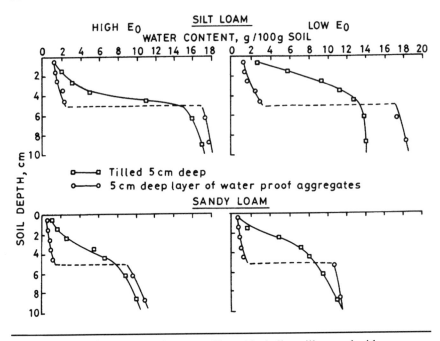

FIGURE 4.7. Near-surface moisture profiles with shallow tillage and with waterproof aggregates at the surface in silt loam and sandy loam soils under evaporativities of 3.6 ± 0.2 and 15.6 ± 0.7 mm day^{-1} (Jalota and Prihar, 1992)

of water only in vapor phase in tilled soil (Hanks, 1958; Hanks and Woodruff, 1958; Acharya and Prihar, 1969) were far from realistic. With postwetting tillage, capillaries were broken only partially, and the tilled layer dried out from the top downward at rates depending upon the water transmission properties of the soil and E_0 (Prihar and Jalota, 1988). Water movement occurred in both liquid and vapor phases at the interface of tilled and untilled layers and within the tilled layer. The contribution of liquid flux to total evaporative flux was greater in finer-textured soils and under lower evaporativities. Vapor flux contributed more when soil was of medium texture and E_0 was high.

Depth, Porosity, and Evaporativity Interactions

Jalota (1993) showed that the evaporative flux (q_v) through waterproof soil mulch decreased with increasing thickness of the mulch and increased with increasing E_0 (Figure 4.9). Using Equation 4.1, he found that vapor flux density through soil mulch was highly correlated with mulch thickness (D)

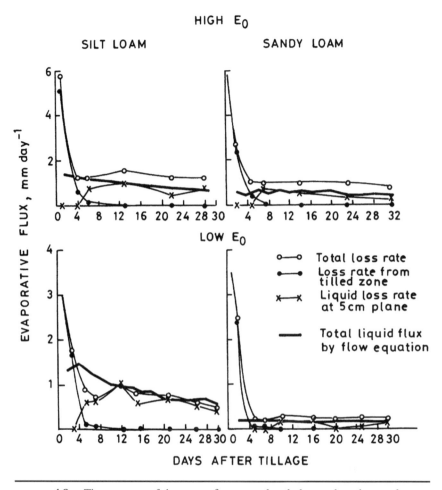

FIGURE 4.8. Time pattern of the rates of computed and observed total water loss and the observed losses from the tilled layer and at 5 cm plane from tilled silt loam and sandy loam soils under evaporativities of 3.6 ± 0.2 and 15.6 ± 0.7 mm day^{-1} (Jalota and Prihar, 1992)

between 5 and 75 mm, mulch porosity (ϕ) between 0.53 and 0.63, wind speed (WS) between 2 and 8 km hr^{-1}, and air temperature (T) between 17°C and 39°C:

$$q_v = 0.0013T^{2.3**} \, WS^{0.29**} \, \phi^{3.95*} \, D^{-0.45**} \qquad (r^2 = .88) \qquad (4.1)$$

where ** and * denote significance at 1% and 5% probability, respectively.

TABLE 4.3   Components of Evaporation Loss After 30 Days from Two Soils Under Two
Evaporativities (E_0)

		Liquid flux by mass balance (%)			
E_0 (mm day^{-1})	Hank's vapor flux	From tilled layer	Across tilled layer	Total	Calculated liquid flux (%)
Silt loam:					
15.6	38	20	42	62	84
3.6	2	26	72	98	97
Sandy loam:					
15.6	51	18	31	49	53
3.6	30	50	20	70	66

Source: Jalota and Prihar, 1992.

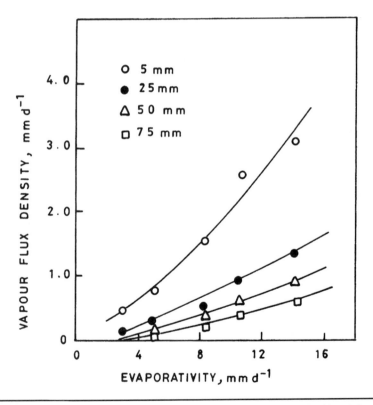

FIGURE 4.9.   Vapor flux density as affected by E_0 and thickness of soil mulch
(Jalota, 1993)

TABLE 4.4 Calculated Vapor Flux Density (q_v) as Influenced by Porosity and Thickness of Waterproof Soil Mulch Under Comparable Evaporativities Induced with Wind and Radiation

E_0 (mm day^{-1})	Porosity (m^3 m^{-3})	Thickness of soil mulch (mm)				
		10	20	50	75	100
		q_v (mm day^{-1})				
9.4 (high temp., low wind)	0.53	0.80	0.40	0.30	0.20	0.20
	0.58	1.10	0.60	0.40	0.30	0.20
	0.63	1.50	0.80	0.50	0.40	0.30
9.5 (high wind, low temp.)	0.53	0.20	0.20	0.10	0.10	0.10
	0.58	0.30	0.20	0.20	<0.10	<0.10
	0.63	0.40	0.20	0.20	<0.10	<0.10

Source: Jalota, 1993.

This regression was used to compute q_v for an E_0 of 9.4 mm day^{-1} achieved with low wind and high radiation and an E_0 of 9.5 mm day^{-1} achieved with high wind and low radiation. The results are given in Table 4.4.

The computed q_v increased with increasing porosity and decreased with increasing thickness of the simulated (dried) tilled layer. The mulch porosity and thickness exhibited a strong interaction with variously induced E_0. For example, a vapor flux density of 0.3 mm day^{-1} at porosities of 0.53, 0.58, and 0.63 m^3 m^{-3} was obtained with 50, 75, and 100 mm thick mulch when an E_0 of 9.4 mm day^{-1} was radiation induced. But for a wind-induced E_0 of 9.5 mm day^{-1}, the same q_v for a porosity of 0.58 was obtained with only 10 mm thick mulch.

Thus, for a given E_0, a shallower tilled layer of dry soil could be as effective in reducing evaporation losses under high wind velocity as a several times thicker layer under high radiation.

Effect of Tillage and Soil Type on the Moisture Profile

Like the cumulative evaporation the depth distribution of water in the soil profile is also influenced by soil type, E_0, and tillage. Gill et al. (1977), Gill and Prihar (1983), and Jalota (1984) reported that after a few weeks of drying the soil water content profiles in untilled soil were continuous with depth, and the water content gradients were larger in the coarser-textured soils and under higher E_0. These observations support the earlier theoretical considerations of Covey (1963). The tilled layer above the interface dried out faster than the corresponding depth of the untilled soil for reasons mentioned earlier. This phenomenon was more pronounced under higher E_0. The rate of

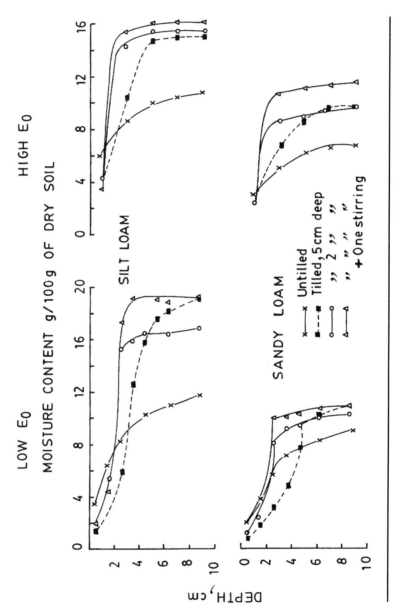

FIGURE **4.10.** Near-surface soil moisture profiles at 30 days of drying as affected by depth and frequency of tillage, soil type, and E_0 (Jalota and Prihar, 1990a)

advance of the isohydral front in the tilled zone was relatively more rapid until it reached the bottom of the tilled layer. After reaching the junction, its rate of advance was markedly reduced, irrespective of E_0 (Gill and Prihar, 1983). However, the reduced rate of advance persisted for a longer period with deeper tillage and a decrease in E_0.

The undisturbed subsoil below the tilled layer dried more slowly than the corresponding layers of the untilled soil (Hillel and Hadas, 1972) because of the interface resistance to water flow in a structurally layered soil. In the tilled soil more moisture was retained in the layers immediately below the tilled zone (Sandoval and Benz, 1966; Prihar, Singh, and Sandhu, 1968).

Water content profile was further affected by depth of tillage and additional stirring of the tilled layer. For example, Jalota and Prihar (1990a) reported that in silt loam drying under an E_0 of 4.6 mm day^{-1}, the residual soil water content (by weight) in the 3–4 cm layer 30 days after tillage averaged 9% and 15% with 5 cm and 2 cm deep tillage and 18% when the 2 cm tilled layer was stirred again after 1 day (Figure 4.10). In the sandy loam under the same conditions, the corresponding soil water contents were 3% and 8% with 5 cm and 2 cm deep tillage and 10% for the 2 cm tilled layer stirred again. Similarly, under an E_0 of 15.6 mm day^{-1}, the soil water content of this layer averaged 10%, 14%, and 16% in silt loam, and 5%, 8%, and 11% in sandy loam with 5 cm deep, 2 cm deep, and 2 cm deep tillage plus stirring, respectively.

Summary

Early in the twentieth century dust mulch created by shallow tillage was advocated for conserving water by reducing evaporation from soil. But its actual effect remained a matter of controversy for a long time. More recent studies, the subject matter of this chapter, have increased our understanding of the time pattern of water loss from untilled and tilled soil. Shallow tillage of the surface layer alters the albedo of soil, the surface area exposed to the atmosphere, and the penetration of wind into soil. All these changes accelerate the rate of water loss and aid the development of a dry layer, which tends to reduce further water loss from the tilled soil vis-à-vis the untilled soil. Eventually, a dry layer develops at the surface of the untilled soil too, which reduces the rate of evaporation, often below that from the tilled soil. Hence, the benefit of tillage in evaporation reduction is time variant; evaporation reduction increases for some time, peaks at a point, and decreases thereafter.

The time course and magnitude of evaporation reduction by tillage has been shown to depend upon the water characteristics of the soil, atmospheric evaporativity (E_0), time (after wetting), and type of tillage (porosity, depth, and clod size distribution of the tilled layer). The benefit is greater and lasts longer with fine-textured soils under low E_0. With self-mulching coarse-textured soils and under high E_0, the benefit is small and may sometimes be negative. Between these two extremes there are numerous soil, E_0, and tillage (time and type) combinations where tillage reduces evaporation losses to varying degrees.

The earlier assumptions of the complete destruction of capillarity and of the movement of water only in vapor phase in the tilled layer have been found unrealistic. The locus of evaporation in tilled soil does not instantaneously shift to the bottom of the tilled layer. In fact, the upward liquid flow through the junction of tilled and untilled soil continues for a variable length of time depending upon soil and climatic conditions. Additional post-tillage stirring of the (shallow) tilled layer further accelerates formation of a dry layer, thus retarding the upward liquid flow. Stirring was found to enhance the benefit of tillage for evaporation reduction in fine-textured soils. Interestingly, the effectiveness of a completely dry layer in reducing vapor loss under a given E_0 depended on the components of E_0. For example, under wind-induced E_0, a relatively much shallower layer reduced vapor loss to the same extent as a much thicker layer of the same porosity under radiation-induced E_0.

References

Acharya, C. L., and S. S. Prihar. 1969. Vapor losses through soil mulch at different wind velocities. *Agronomy Journal* 61:666–668.

Allmaras, R. R. 1967. Soil water storage as affected by infiltration and evaporation in relation to tillage-induced soil structure. In *Tillage for Greater Crop Production*, Conference Proceedings of the American Society of Agricultural Engineering. Saint Joseph, Mich.

Aujla, T. S., and S. S.Cheema. 1983. Modifying profile water storage through tillage, herbicide, chemical evaporation retardant, and straw mulch and its effect on rainfed chick pea (*Cicer arietinum* L.). *Soil and Tillage Research* 3:159–170.

Baver, L. D. 1966. *Soil Physics*. Bombay: Asia Publishing House.

Benoit, G. R., and D. Kirkham. 1963. The effect of soil surface conditions on evaporation of soil water. *Soil Science Society of America Proceedings* 27:495–498.

Call, L. E., and M. C. Sewell. 1917. Soil mulch. *Journal of the American Society of Agronomy* 9:49–61.

Chaudhary, R. S., and C. L. Acharya. 1993. A comparison of evaporative losses from soil of different tilths and under mulch after harvest of rice. *Soil and Tillage Research* 28:191–199.

Covey, W. 1963. Mathematical study of first stage of drying of moist soil. *Soil Science Society of America Proceedings* 27:130–134.

Ehlers, W., and R. R. van Der Ploeg. 1976. Evaporation, drainage, and unsaturated hydraulic conductivity of tilled and untilled fallow soil. *Zeitschrift für Pflanzenernährung und Bodenkunde* 139:373–376.

Eser, C. 1884. Untersuchungen über den Eingluss der physikalischen und chemischen Eigenschaften des Bodens aufdessen Verdunstungsvernegen. *Forschung Gebiet Landwirtschaftlich Physik* 7:1–124.

Gardner, W. R., and M. Fireman. 1958. Laboratory studies of evaporation from soil columns in the presence of a water table. *Soil Science* 85:244–249.

Gill, K. S., S. K. Jalota, S. S. Prihar, and T. N. Chaudhary. 1977. Water conservation by soil mulch in relation to soil type, time of tillage, tilth, and evaporativity. *Journal of the Indian Society of Soil Science* 25:360–366.

Gill, K. S., and S. S. Prihar. 1983. Cultivation and evaporativity effects on drying patterns in sandy loam soil. *Soil Science* 135:366–377.

Hanks, R. J. 1958. Vapor transfer in dry soil. *Soil Science Society of America Proceedings* 22:372–374.

Hanks, R. J., and W. R. Gardner. 1965. Influence of different diffusivity water content relations on evaporation of water from soil. *Soil Science Society of America Proceedings* 29:495–498.

Hanks, R. J., and N. P. Woodruff. 1958. Influence of wind on vapor transfer through soil, gravel, and straw mulches. *Soil Science* 86:160–164.

Heinonen, R. 1985. *Soil Management and Crop Water Supply.* 4th ed. Department of Soil Science, Swedish University Agricultural Science, S-75007. Uppsala, Sweden: Soveringes Lantbruksuniversitet.

Hillel, D., and A. Hadas. 1972. Isothermal drying of structurally layered soil columns. *Soil Science* 113:495–498.

Hillel, D., C. H. M. van Bavel, and H. Talpaz. 1975. Dynamic simulation of water storage in fallow soil as affected by mulch of hydrophobic aggregates. *Soil Science Society of America Proceedings* 39:826–833.

Holmes, J. W., E. L. Greacen, and G. C. Gurr. 1960. The evaporation of water from bare soils with different tilths. In *Transactions of the 7th International Congress of Soil Science*, vol. 1, pp. 188–194. Wageningen, the Netherlands: International Society of Soil Science.

Jacks, G. V., W. D. Brind, and R. Smith. 1955. *Mulching.* Bureau of Soil Technical Communication no. 49. Farnham Royal, Bucks, England: Commonwealth Agricultural Bureaux.

Jalota, S. K. 1984. Drying pattern of bare and tilled silt loam, sandy loam, and loamy sand soils as affected by zero time water profiles and evaporativity. Ph.D. diss., Punjab Agricultural University, Ludhiana, India.

_____. 1993. Evaporation through soil mulch in relation to mulch characterisics and evaporativity. *Australian Journal of Soil Research* 31:131–136.

_____. 1995. Estimating soil evaporation parameters in relation to soil, tillage depth, and evaporativity. *Journal of Indian Society of Soil Science* 43:667–668.

Jalota, S. K., and S. S. Prihar. 1979. Soil water storage and weed growth as affected by shallow tillage and straw mulching with and without herbicide in barefallow. *Indian Journal of Ecology* 5:41–48.

_____. 1990a. Bare soil evaporation in relation to tillage. *Advances in Soil Science* 12:187–216.

_____. 1990b. Effectiveness of tillage-induced soil mulch as influenced by soil type and atmospheric evaporativity. *Journal of the Indian Society of Soil Science* 38:583–591.

_____. 1992. Liquid component of evaporative flow in two tilled soils. *Soil Science Society of America Journal* 56:1893–1898.

Minhas, P. S., B. K. Khosla, and S. S. Prihar. 1986. Evaporation and redistribution of salts in a silt loam soil as affected by tillage-induced soil mulch. *Soil and Tillage Research* 7:301–313.

Papendick, R. I., M. J. Lindstrom, and V. Cochran. 1973. Soil mulch effects on seedbed temperature and water during fallow in eastern Washington. *Soil Science Society of America Proceedings* 37:307–314.

Penman, H. L. 1941. Laboratory experiments on evaporation from fallow soil. *Journal of Agricultural Science* 31:454–463.

Prihar, S. S., and S. K. Jalota. 1988. Role of shallow tillage in soil water management. In *Proceedings of the International Conference "Dry Land Farming: A Global Perspective,"* edited by P. W. Unger, T. V. Sneed, W. R. Jorden, and R. L. Jensen, pp. 128–130. Amarillo/Bushland: Texas Agricultural Experiment Station.

Prihar, S. S., B. Singh, and B. S. Sandhu. 1968. Influence of soil and environment on evaporation losses from mulched or unmulched pots. *Journal of Research, Punjab Agricultural University, Ludhiana,* 5:320–328.

Prihar, S. S., and D. M. van Doren Jr. 1967. Mode of response of weed-free corn to post-planting cultivation. *Agronomy Journal* 59:513–516.

Ramacharlu, P. T. 1957. Rate of evaporation of water in relation to particle distribution in soils. *Journal of the Indian Society of Soil Science* 5:117–121.

Sandoval, F. M., and L. C. Benz. 1966. Effect of bare fallow, barley, and grass on salinity of a soil cover over a saline water table. *Soil Science Society of America Proceedings* 30:392–396.

Scotter, D. R., and P. A. C. Raats. 1969. Dispersion of water vapor in soil due to air turbulence. *Soil Science* 108:170–176.

Singh, D., and S. D. Nijhawan. 1944. A critical study of the effect of soil mulch on evaporation of soil moisture. *Indian Journal of Agricultural Sciences* 14:127–137.

Unger, P. W. 1984. *Tillage Systems for Soil and Water Conservation.* FAO Soils Bulletin no. 54. Rome, Italy.

Veihmeyer, F. J. 1927. Some factors affecting irrigation requirements. *Hilgardia* 2:125–291.

Wadham, L. A. 1944. The flow of heat through granular materials. *Journal of the Chemical Society of India* 63:337–343.

Widstoe, J. A. 1909. *Irrigation Investigation.* Utah Agricultural Experimental Station Bulletin 105. Logan.

Wiese, A. F., and T. J. Army. 1958. Effect of tillage and chemical weed control on soil moisture storage and losses. *Agronomy Journal* 50:465–467.

Willis, W. O., and J. J. Bond. 1971. Soil water evaporation: Reduction by simulated tillage. *Soil Science Society of America Proceedings* 35:526–528.

5

Evaporation Reduction by Straw Mulching

The usual effect of crop residue is to lower the maximum soil temperature because of the residue's greater reflection and less absorption of solar radiation and lower thermal conductivity than soil (Hanks, Bowers, and Bark, 1961; Lemon, 1956; Lehane, 1961; McCalla and Duley, 1946; McCalla, 1947; van Doren and Allmaras, 1978). The reflection effect of radiation on soil temperature approaches a maximum as surface residue coverage approaches 100% (van Doren and Allmaras, 1978). When the amount of residue exceeds 100% coverage, the soil temperature is reduced because of the insulating effect of the mulch. For example, 8–12 Mg ha^{-1} of wheat (*Triticum aestivum* L.) straw at the surface resulted in lower soil temperature during a hot period and higher temperature during a cold period than 4 Mg ha^{-1} of wheat straw, an amount that provided almost 100% coverage (Unger, 1978). The insulation effect increased with increasing thickness of the residue layer (Gupta and Gupta, 1983).

A reduction of soil temperature with residues at the surface has been widely reported (Aston and Fischer, 1986; Black and Siddoway, 1979; Carter and Rennie, 1985; Wall and Stobbe, 1984), but the magnitude of reduction varied with the rate of the residue and the prevailing weather conditions. In north India, application of wheat straw mulch at a rate of 6 Mg ha^{-1} immediately after sowing of corn in June reduced the maximum soil temperature at 5 cm depth from 45.7°C to 36°C and the amplitude from 14.5° to 9.5°C (Singh and Sandhu, 1979). Similarly, the monolayer mats of typha (*Typha latifolia* L.) decreased the maximum soil temperature at 10 cm depth by 11°C in the drier moisture regime and by 1–2°C immediately following wetting (Bansal, Gajri, and Prihar, 1971). Smika (1983) showed that apart from the amount of residue

TABLE 5.1   Average daily maximum soil surface
temperature in Akron, Colorado, during 5 weeks
in August and September as affected by
positioning of wheat straw

Straw position	Soil temperature (°C)
Bare soil	47.8c
Flat straw	41.7b
¾ flat, ¼ standing	39.6b
½ flat, ½ standing	32.2a

Source: Smika, 1983.
Note: Any two means not followed by the same letters are
significantly different at the 5% level.

per se, the effect of mulch on soil temperature depended upon the fraction of residue lying flat at the surface and the fraction standing as stubble (Table 5.1).

A lower temperature at the soil surface underneath the residue lowers the vapor pressure and, thus, the vapor pressure gradient between the soil surface and the atmosphere above it. It also decreases the rate of vapor transfer from soil to the atmosphere by increasing resistance to vapor flow.

General Trends

The time trends of evaporation rates from straw-mulched and bare soil discussed in Chapter 2 indicate that evaporation reduction with mulch is a quadratic function of time and depends upon the dynamics of the evaporation rate from the mulched soil relative to that from the bare soil. Overall reduction in evaporation loss with mulch is influenced by soil type, evaporativity (E_0), nature and amount of residue, timing and manner of mulch placement, precipitation pattern, and other climatic factors and tillage practices. Therefore, it is important to manage crop residues for evaporation reduction in relation to these variables.

Effect of Soil Type

Crop residue placed at the soil surface reduces evaporation as long as the soil surface remains wet. It is well known that in fine-textured soils of high water retentivity the surface layer remains wet for longer periods than in coarse-textured soils. Therefore, the total evaporation reduction with residue mulch is greater and longer lasting in finer-textured soils, which also maintain higher hydraulic conductivity in the drier range (Wind, 1961).

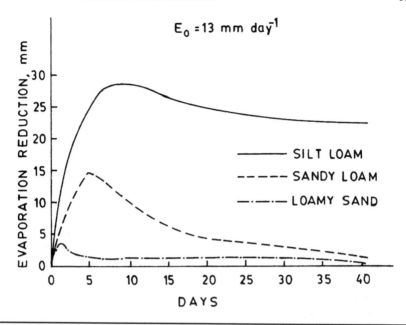

FIGURE 5.1. Time course of evaporation reduction from initially wet, 95 cm deep soil columns of silt loam, sandy loam, and loamy sand as affected by 3 Mg ha^{-1} wheat straw mulch under an E_0 of 13.0 mm day^{-1}

A field study by Singh and Nijhawan (1944) showed that evaporation reduction with pearl millet (*Pennisetum typhoides* L.) residue mulch was greater on a clay loam soil than on a sand soil. This has been corroborated by the evaporation simulation models of Saxton et al. (1988). Controlled studies in 1994 by Jalota (unpublished) have also shown that evaporation reduction with wheat straw mulch was much greater and extended over a longer period in silt loam soil than in loamy sand soil under an E_0 of 13.0 mm day^{-1}. The behavior of sandy loam soil was intermediate (Figure 5.1) to the other two soils.

Effect of Evaporativity

Atmospheric evaporativity affects evaporation reduction by residue cover through its effect on relative evaporation rates from bare and residue-covered soils. Under high E_0, evaporation reduction with straw mulch is greater than under low E_0 and continues as long as the evaporation rate from the residue-covered soil remains less than that from the bare soil. Thereafter, the benefit of evaporation reduction declines when the evaporation rate from bare soil decreases because of the intense drying of the surface layer under high E_0. Under low E_0, because of the slow rate of drying the cumulative evaporation reduction increases more gradually, but it continues to increase for a longer period (Jalota and Prihar, 1990).

Effect of the Interaction between Soil Type and Evaporativity

Evaporation reduction by mulching exhibits a strong interaction between soil type and evaporativity. On soils of low water retentivity, which self-mulch rapidly under high E_0, residue mulching may have little benefit. On soils of high retentivity, where the dry layer at the surface develops more slowly, mulching does reduce evaporation. Prihar, Singh, and Sandhu (1968) reported from a pot study that during summer the advantage of wheat straw mulching for evaporation reduction was greater on fine-textured than on coarse-textured soil. This has also been supported by a 1994 simulation study by Jalota (unpublished). Under an E_0 of 16 mm day^{-1}, mulching with 6 Mg ha^{-1} of wheat straw reduced cumulative evaporation after 30 days of drying by 50, 6, and 7 mm on silt loam, sandy loam, and loamy sand soils, respectively. However, under an E_0 of 2 mm day^{-1}, the respective reductions at 30 days were only 6, 6, and 4 mm. The cumulative evaporation reduction is likely to increase further with time under such a low E_0.

Effect of Mulch Rates

Evaporation reduction with straw mulching has been related to either the percentage of the surface covered by residue (or the amount of residue per unit area) or the thickness of the residue mulch. The percentage cover and thickness of the mulch layer are governed by the mass applied per unit area and the density of the mulching material (Unger and Parker, 1976). For a residue of given density, the fraction of soil covered by a monolayer of mulch increases with the amount of residue per unit area. Once the area is completely covered, further additions increase the thickness of the mulch layer.

Based on experimental data collected by various techniques, Soleneker and Moldenhauer (1977), Wischmeier and Smith (1978), and Gregory (1982) attempted to relate the fraction of (soil) surface covered to the mass of residue applied. According to Gregory (1982), the fraction of soil surface covered (F_c) increased with the increase in dry mass of the residue mulch as follows:

$$F_c = 1 - \exp(-A_m M) \qquad (5.1)$$

TABLE 5.2   Values of the coefficient A_m (area covered per unit residue) for different crop residues obtained from data in the literature using Equation 5.1

Residue	A_m (ha kg^{-1})	Based on data from:
Oats	0.0014	Soleneker and Moldenhauer, 1977
Soybeans	0.00072	Soleneker and Moldenhauer, 1977
Wheat	0.00054	Wischmeier and Smith, 1978
Wheat	0.00045	Gregory, 1982
Corn	0.00040	Soleneker and Moldenhauer, 1977

Source: Gregory, 1982.

where A_m is the area covered per unit mass of residue (ha kg^{-1}), and M is the mass of residue applied per unit area (kg ha^{-1}). The values of A_m obtained for residues of different crops are given in Table 5.2.

Whereas the fraction of area covered governs energy interception, the thickness or roughness of the residue governs vapor diffusion. Jalota and Prihar (1990) attempted to separate the "energy interception" and "resistance to vapor flow" components of the mulch effect in reducing constant-rate stage evaporation. Reduction (R) in the constant-rate stage evaporation rate was expressed as $R = 1 - E_m/E_0$ where E_m is the constant-rate stage evaporation from the mulched soil. They plotted R as a function of (wheat straw) mulch rates. The linear tail portion of this curve, which represented the resistance to vapor flow, was extrapolated. This effect was then deducted from the total R to get the energy interception component (Figure 5.2). Evaporation reduction due to energy interception increased with mulch rate up to 9 Mg ha^{-1}, beyond which the effect was solely attributable to regulation of vapor flow.

Evaporation reduction from wet soil has been reported to increase linearly with percentage of cover (Willis, 1962). Partial covers obtained with a given amount of mulch are most effective when the soil is wet, and their efficiency can be improved further by increasing the area of continuous cover. In other words, if a cover is cut into strips, the effectiveness of the cover would decrease even though the total percentage of the area covered is not changed. Smaller partial covers are ineffective in reducing evaporation losses especially when the E_0 is high. Under such situations, greater surface residue rates are necessary to reduce evaporation during initial periods of drying, that is, during the constant-rate stage (Bond and Willis, 1969).

Under a given E_0, the constant-rate stage evaporation rate from residue-covered Pullman clay loam was related to residue thickness (Unger and Parker, 1976):

$$\log E_m = a + bM' \tag{5.2}$$

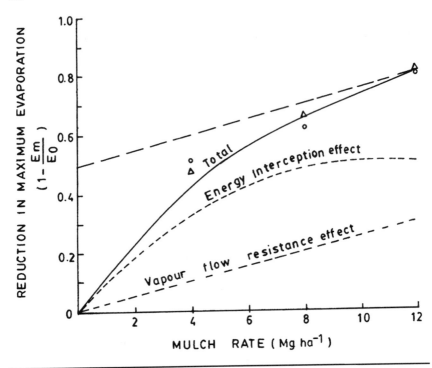

FIGURE 5.2. Separation of straw mulch effects on reduction of constant-rate stage evaporation from soil into energy interception and resistance to vapor flow components (Jalota and Prihar, 1990)

where E_m is the constant-rate stage evaporation rate (cm day^{-1}) on mulched soil, M' is the thickness of mulch (cm), and a and b are constants. Jalota and Prihar (1990) reported that during the constant-rate stage of drying, the evaporation rate from (wheat) straw-mulched sandy loam soil was related to mulch rate and E_0 as follows:

$$E_m = E_0 \exp(-0.135M) \qquad (r^2 = .89, n = 6) \tag{5.3}$$

where E_m is the constant-rate stage evaporation rate (mm day^{-1}) from mulch-covered soil, E_0 is the evaporativity (mm day^{-1}), and M is the amount of mulch (Mg ha^{-1}). The reduction in evaporation rate per unit of mulch was larger at low mulch rates and decreased with increasing amounts of mulch. Steiner (1989) used Unger and Parker's (1976) data to relate constant-rate stage evaporation from mulched soil (E_m) with the depth of mulch (m^3 m^{-2}). The latter was obtained by dividing the amount of residue per unit surface

TABLE 5.3   Values of coefficients for the best fit relation of the ratio of energy-limited evaporation rate from mulched pots to free water evaporation (E_m/E_0) as a function of residue rate and evaporativity

Soil type	a	b	c	r^2	n
Silt loam	1.061	−0.218	−0.0221	.95	16
Sandy loam	1.040	−0.225	−0.0074	.99	16
Loamy sand	1.223	−0.224	−0.0388	.90	12

Source: Prihar, Jalota, and Steiner, 1996.

Note: $E_m/E_0 = a \exp[-(b\text{RES} + cE_0)]$. RES is in Mg ha^{-1}; E_0 is in mm day^{-1}.

(Mg m^{-2}) by the crop-specific density (Mg m^{-3}) of the residue for cotton (*Gossypium hirsutum* L.), sorghum (*Sorghum bicolor* L. (Monech)), and wheat individually. This thickness was much smaller than the measured thickness because the solid matter in the residue was assumed to be uniformly distributed flat on the soil surface. The different crop residues were still similarly effective in controlling the relative evaporation rate during constant-rate stage evaporation. On pooling the data for all three types of crop residues, she found the following relation gave the best fit under a given E_0:

$$E_m = -0.99 - 0.236(\ln M'') \tag{5.4}$$

where M'' is the mulch rate (m^3 m^{-2}).

From later experiments involving more than one E_0, Steiner, Jalota, and Prihar (1990) found that on sandy soils residue level alone was not adequate to predict relative evaporation rate and a model including E_0 (Equation 5.5) was needed for better prediction of evaporation rate:

$$E_m/E_0 = a + b[\ln(\text{RES})] + cE_0 \tag{5.5}$$

where RES is the rate of wheat residue (Mg ha^{-1}), and a, b, and c are constants. However, this relation exhibits a discrepancy inasmuch as the value of predicted E_m/E_0 for RES = 0 exceeds unity, which is physically untenable. To avoid this discrepancy Prihar, Jalota, and Steiner (1996) incorporated the evaporativity term in Jalota and Prihar's (1990) relation to get Equation 5.6 and determined the values of coefficients for different soils (Table 5.3):

$$E_m/E_0 = a \exp[-(b\text{RES} + cE_0)] \tag{5.6}$$

Equations 5.3 and 5.6 also reflect that evaporation reduction per unit residue is greater at lower mulch rates. Sharma, Upadhay, and Tomar (1985) reported that on a Vertisol, per unit efficiency of mulch for reducing evaporation was greater with mulch (wheat straw) rates less than 4.48 Mg ha^{-1}.

Effect of Type of Residue

The type of residue affects evaporation through the residue's textural characteristics, which govern rates of vapor diffusion through the material. Unger and Parker (1976) compared the effectiveness of wheat, sorghum, and cotton residues in reducing evaporation. A given mass of wheat straw placed on the soil surface was about twice as effective in decreasing evaporation as the sorghum stubble and more than four times as effective as the cotton stalks. The differences were attributed primarily to the physical nature of the residues (hollow, pithy, or woody), which determined their specific densities and hence the thickness of the mulch layer and the surface coverage. Based on rank analysis, the importance of different parameters of the residue for evaporation reduction followed the order:

residue thickness > surface cover > residue application rate > potential evaporation > residue specific density > relative humidity

It was estimated that to achieve an evaporation reduction comparable to that with 8 Mg ha^{-1} of wheat residue, it was necessary to apply 2- and 32-fold larger amounts of sorghum and cotton residues, respectively, in the Great Plains.

Likewise, Greb (1967) reported that assuming perfect distribution, the amounts of residue needed to form a monolayer over 100% of an area were 3600, 2300, 3600, 8100, 6800, and 16,400 kg ha^{-1} for winter wheat, spring barley (*Hordeum vulgare* L.), spring oats (*Avena sativa* L.), German millet (*Panicum millaceum* L.), Sudan grass (*Sorghum sudanese* L.), and sorghum, respectively. The higher efficiency of barley residue was attributed to its thinner stems.

Turkey and Schoff (1963) studied the influence of different mulching materials on soil environment with 15 cm thick mulch on a silt loam soil for five years. They used decomposable (groundnut [*Arachis hypogrea* L.] husk, corncobs [*Zea mays* L.], and sawdust) and nondecomposable (granular foam, rubber, fiberglass, and gravel) materials as mulch. Since there were no significant differences between decomposable and nondecomposable materials, it indicated that the effects of mulch on evaporation were caused mainly by the physical changes in the soil environment.

Effect of Placement of Residue

Evaporation reduction with residues placed at the soil surface varies with soil type, time of application of residue after wetting, and prevailing weather conditions (Prihar, Jalota, and Steiner, 1996; Bond and Willis, 1969; Jalota, 1990). During the constant-rate stage of drying, covering the maximum surface with a residue is most effective. The advantage from surface-applied residue is greater under low E_0 and on fine-textured soils (Prihar, Singh, and Sandhu, 1968).

Recent advances in mechanization have made it possible to mix the residues into the top few centimeters of soil or place it in the subsurface. Residue placed in the subsurface reduces water loss by interrupting much of the capillary movement of water between the two soil layers, besides its insulating effect on the soil layer below it. Unger and Parker (1968) studied the effect of residue over repeated cycles of wetting and drying and found that cumulative evaporation with residue placed at and 15 cm below the surface was 57% and 19% less, respectively, than with residue mixed into the surface layer. Higher evaporation when residue is mixed into the surface layer or placed below the surface occurred because water added to the soil was largely retained in the upper layer and was lost rapidly to the atmosphere. When straw was at the surface, water added to the soil rapidly entered the soil and was lost to the atmosphere through vapor diffusion from the soil surface.

Minhas and Gill (1985) studied the effect of different residue management treatments on evaporation loss from two fine-textured soils. Compared with bare soil, 7 Mg ha^{-1} of oat straw placed 5 cm below the surface reduced cumulative evaporation at 62 days (after five drying cycles) by 42%; straw placed at the surface resulted in a 33% reduction; and straw mixed into the top 0–5 cm resulted in a 28% reduction.

The apparent anomaly in the efficiency of subsurface placement of straw in reducing evaporation (i.e., the discrepancy between the results of Unger and Parker and of Minhas and Gill) may be attributed to the difference in depth of residue placement in the two cases. With the residue placed at a shallow depth (Minhas and Gill, 1985), a small proportion of the applied water was retained in the thin soil layer above the residue, and a major part moved into deeper layers across the residue layer, where it was less susceptible to evaporation. But when residue was placed at a deeper depth (Unger and Parker, 1968), most of the applied water was retained in the thick soil layer above the residue and was easily lost to the atmosphere.

Effect of Rainfall

From a comprehensive review of literature on mulching, Jacks, Brind, and Smith (1955) concluded that mulching with crop residues reduced evaporation only where the soil surface was maintained at a high moisture content by frequent rains or a high water table. Russel (1939) and Hanks and Woodruff (1958) also suggested that mulches conserved moisture efficiently during frequent rains but decreased in value during prolonged dry periods. Bond and Willis (1969) suggested a strong interaction between rainfall patterns and amount of residue. In years of above-normal precipitation, straw mulch material spread on the entire surface resulted in greater evaporation reduction. But when precipitation was below normal and evaporativity was high and surface residue deteriorated, the potential for decreasing soil water evaporation by residue from small-grain crops was enhanced by placing it on less than the entire surface.

Jalota and Prihar (1990) reported the time trends of evaporation reduction with residue mulching in the absence of precipitation under controlled laboratory conditions. Cumulative evaporation reduction generally increased with time up to a certain period and decreased thereafter depending upon E_0. Under an E_0 of 3.3 mm day^{-1}, the reduction in evaporation with 4 Mg ha^{-1} of wheat straw continued to increase for 35 days and beyond after the application of mulch. However, for evaporativities of 10.2 and 15.4 mm day^{-1}, the cumulative reduction under a given mulch rate peaked much earlier and decreased thereafter (Figure 5.3). For a given E_0, the higher the mulch rate, the greater the evaporation reduction and the later the cumulative reduction peaked.

The differential behavior of evaporation reduction with mulching under low and high evaporativities was attributed to time trends of evaporation rates from mulched and bare soil under different evaporativities (Jalota and Prihar, 1990). Under an E_0 of 3.3 mm day^{-1}, the evaporation rate from soil with 4 Mg ha^{-1} of mulch remained lower than that from bare soil throughout the period of study. Under an E_0 of 10.2 mm day^{-1}, this trend continued for 8, 14, and 20 days with mulch rates of 4, 8, and 12 Mg ha^{-1}, respectively (Figure 5.4). Similarly, under an E_0 of 15.4 mm day^{-1}, the evaporation rate

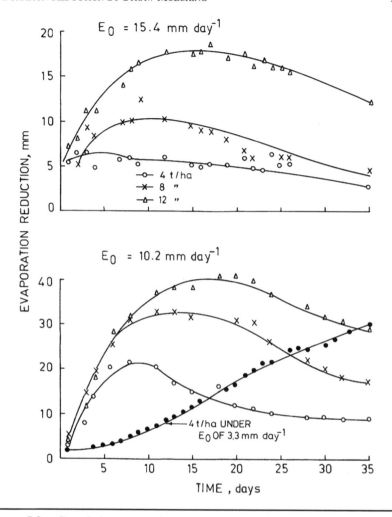

FIGURE 5.3. Cumulative evaporation reduction as affected by straw mulch rates and evaporativity in a sandy loam soil (Jalota and Prihar, 1990)

from bare soil exceeded that from the mulched soil for 3, 7, and 15 days for 4, 8, and 12 Mg ha^{-1}, respectively. Thereafter, the rate of water loss from the mulched soil exceeded that from the bare, and hence the cumulative evaporation reduction started declining. For a given mulch rate, the maximum evaporation reduction was lower and occurred earlier under higher than under lower E_0.

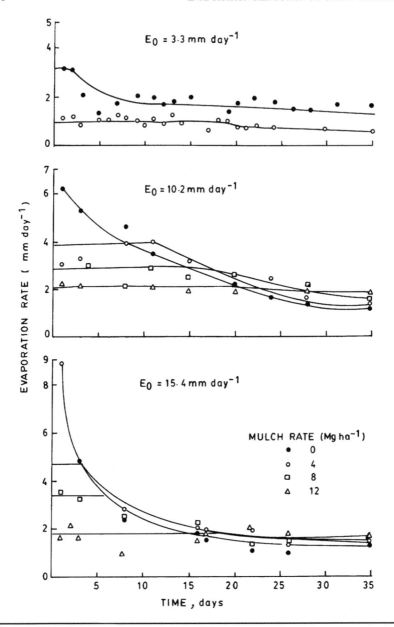

FIGURE 5.4.   Effect of straw mulch rates on time trends of evaporation from a sandy loam soil under three evaporativities (Jalota and Prihar, 1990)

Field Observations

Direct and precise measurement of evaporation from residue-covered soil under field conditions is difficult because of interacting factors, such as residue orientation (flat or standing, which affects porosity and thickness of the layer), nonuniformity of the residue layer, rainfall interception by residue, and aerodynamic roughness. Many researchers (Unger, 1978, 1984; Lal, DeVleeschauwer, and Nganje, 1980; Hundal and De Datta, 1982; Ojeniyi, 1986; Rasmussen, Newhall, and Cartee, 1986; al-Darby et al., 1989; Nyborg and Malhi, 1989) have studied the integrated effects of mulching and tillage practices on soil water storage under field conditions at various locations. The soil water storage increased with increasing amounts of residues at the surface. Field studies in Texas, United States, showed that application of 1 Mg ha^{-1} of wheat residue during an 11-month fallow period increased water storage by 36 mm. Application of 4 and 8 Mg ha^{-1} increased it further by an additional 12 and 10 mm, respectively (Unger 1978). However, these figures do not reflect the effect of evaporation alone but also the effect on infiltration and redistribution of water. Greb (1983) summarized the results from four locations in the Great Plains, United States, on soil water gain during fallow for different rates of surface mulches of small-grain crop residues. In most cases the cropping system was alternate wheat-fallow. These long-term experiments showed that increases in soil water with mulches were significant and proportional to the amount of residue, which ranged from 2.2 to 6.6 Mg ha^{-1}; the storage efficiency per unit mulch decreased slightly with an increase in mulch rate. It was further observed that storage efficiency was generally higher in the lower-temperature regions.

Residues at the surface reduce evaporation by reducing wind speed and soil temperature (Smika, 1983; Enz, Brun, and Larson, 1988). Smika (1983) also showed that a given amount of residue on the soil surface caused differential evaporation loss because of the difference in orientation of the residue. As the fraction of standing residue increased, greater wind speed was needed to initiate water loss. In addition, the rate of water loss at a given wind speed decreased with an increase in the fraction of residue standing as stubble (Table 5.4).

In India, Singh and Nijhawan (1944) compared the moisture-conserving effects of mulching with pearl millet (*Pennisetum typhoides* L.) stalks and tillage-induced soil mulch. Mulching with millet stalk was more effective over shorter periods, and soil mulch was more effective over longer periods. The higher initial benefit of the residue mulch was attributed to the fact that 39%–46% of the water loss had occurred before the soil could be cultivated to form soil mulch. But after the soil mulch was formed, the rate of loss was

TABLE 5.4   Soil water loss as affected by bare soil and
straw position during 5 weeks in August and September

Straw position	Soil water loss (mm)	
	Daily average	Total
Bare soil	0.66c	23.1f
Flat straw	0.56b	19.6e
¾ flat, ¼ standing	0.53b	18.6e
½ flat, ½ standing	0.43a	15.1d

Source: Smika, 1983.
Note: Any two values not followed by the same letters are significantly different at 0.01 probability.

less under soil mulch than under residue mulch. Aujla and Cheema (1983) observed that at 46 days after application of 6 Mg ha^{-1} of wheat straw, the water storage in a 0–180 cm profile was 32 mm higher under straw mulch than under the bare soil.

Jalota and Prihar (1979) showed that the water conservation effects of straw mulch are influenced by season. Mulching a sandy loam soil with 6 Mg ha^{-1} of wheat straw during the high E_0 period of March to June (82 days) reduced evaporation losses from 123 to 101 mm. During the low E_0 months of September to January (123 days), mulching reduced water loss from 58 mm to 54 mm. The rainfall during the high- and low-evaporativity periods was 119 mm and 56 mm, respectively, which occurred in 15 and in 4 storms.

The reduction in evaporation from soil with crop residues is reflected in increased profile water storage during fallow (Greb, Smika, and Black, 1967, 1970; Ali and Prasad, 1972). Several studies (Bansal, Gajri, and Prihar, 1971; Lal, 1974; Khera et al., 1976; Singh and Sandhu, 1979; Unger, 1978, 1986) have shown that straw mulching caused a higher water content in the surface soil layers even under cropped conditions. Borst and Mederski (1957) reported that in a five-year study, the water content in the 0–15 cm soil layer of plots with residue applied in growing corn averaged 1%–9% higher than in the nonmulched plots. Sandhu, Khera, and Singh (1989) reported that compared with no-mulch, mulching with 6 Mg ha^{-1} of rice (*Oryza sativa* L.) straw in summer corn caused a 0.9%–1.7% higher water content in the 0–15 cm layer. The increase with mulching in mentha (*Japanese mint* L.) was 0.5%–2.5%. The total soil water storage in a 0–180 cm soil profile was also higher with mulch than without it; the increase ranged from 0.7 to 5.4 cm for forage maize and from 0.6 to 1.8 cm for summer mung (*Phaseolus aureus* Roxb.).

In a field study, Prihar et al. (1979) observed that freshly cut "basooti" (*Promna mucronata* L.) weed mulch applied at 10 Mg ha^{-1} a month before harvest in standing corn affected water storage favorably in three out of seven years. It increased the 0–180 cm profile storage by 20 mm or more at harvest of the crop. Critical analysis of the rainfall distribution pattern after mulch application indicated that in years when the profile was wetted before or immediately after mulching and there were no rains till harvest of maize, the moisture conservation effect was absent. But when rains were received in the later period after mulching but prior to maize harvest, moisture storage with mulch was significantly higher than without it. However, even in years of negligible effects on total water storage, mulching tended to keep the surface layers more moist for a longer period. Under high E_0 and light rain showers, residue was less efficient because most of the rain water was lost from the mulch itself.

Sharma, Kharwara, and Tewatia (1990) reported that in the hilly area of northern India application of vegetative mulch in standing rainy-season corn (or at its harvest) left more residual moisture in the soil during the postmonsoon period. Sixty days after application of treatments (at corn harvest) soil water storage in the 0–45 cm soil layer averaged 64, 54, and 28 mm higher under a mulch of corn stalks applied at the rate of 10 Mg ha^{-1}, a mulch of sal (*Shorea robusta* L.) leaves applied at the rate of 10 Mg ha^{-1}, and tillage, respectively. Rainfall during this period was 127 mm.

Crop residues have also been reported to increase the efficiency of water use of a number of crops, presumably by suppressing direct evaporation loss from soil. For example, Sharma and Gupta (1990) reported that residues of safflower (*Carthamus finetorius* L.), sorghum cob husk, and soybean (*Glycine max* L.) applied at the rate of 5–6 Mg ha^{-1} increased the water use efficiency of chickpea (*Cicer arietinum* L.), linseed (*Linum ustatissimum* L.), wheat, and safflower by 15%–26%, depending upon seasonal variations. Improvement in the water use efficiency of crops was also observed by Greb, Smika, and Black (1967), Urmani, Pharande, and Quamaurzzman (1973), Sharma et al. (1982), Sharma, Verma, and Gupta (1985), and Sharma and Gupta (1987).

Summary

Evaporation reduction by straw mulching is caused mainly by the resulting physical changes in the soil environment, which alter the supply of energy—a prerequisite for evaporation. Straw mulches reflect more solar radiation than bare soil does, and they provide insulation by retaining air in between the straw elements. Consequently, the temperature at the soil surface is lowered. The reduced temperature underneath the mulch lowers vapor pressure at the soil surface and, thus, the vapor pressure gradient between the soil surface and the atmosphere. Mulch also provides resistance to vapor flow from the soil surface to the atmosphere.

Straw mulching for reducing evaporation is effective as long as the soil remains wet. This condition is met where rains are frequent, E_0 is low, and the rate of replenishment of water from the deeper layers of soil to the surface is high. Under low E_0 and in fine-textured soils, the beneficial effect of straw mulching continues for a long period, but under high E_0 in coarse-textured soils, evaporation reduction increases up to a certain period and declines thereafter. In some cases it even becomes negative. The magnitude and longevity of the beneficial effect are influenced by the amount, type, and method of placement of residue, soil type, E_0, and rainfall pattern. In general, the higher the mulch rate, the greater the evaporation reduction. However, for a given amount of mulch, evaporation reduction may vary with the type of residue. Residues of wheat, barley, and oats have been found to be more effective than those of sorghum or cotton. Apart from the amount and type, the manner of placement of residue, such as spreading at the soil surface, sub-surface placement, and mixing with the surface layer, affects its efficiency for evaporation reduction. Residue placed at a shallow depth was found more effective than deeper placement. In the former case, more of the added water moved below the residue layer, where it was less susceptible to evaporation. Therefore, when attempting to reduce evaporation from soil by means of crop residue, these effects should be taken into consideration.

Field studies with straw mulching have been more extensive, particularly in the context of no-till farming, which leaves crop residues at the soil surface. Profile moisture storage was observed to increase with an increase in the rate of residue per unit area. The effects were larger when rains were more frequent. However, the beneficial effect observed in these studies cannot be entirely attributed to reduction of soil water evaporation by residue mulch. Part of the benefit accrues from the mulch's effect on the infiltrability of the soil. Where rain was followed by long dry spells, the benefit was much lower. Both tillage and straw mulching have been observed to favorably affect the soil moisture regime in growing crops and to increase water use efficiency.

References

Ali, M., and R. Prasad. 1972. Mulching means more moisture. *Indian Farming* 22:38–39.

Aston, A. R., and R. A. Fischer. 1986. The effect of conventional cultivation, direct drilling, and crop residue on soil temperature during the early growth of wheat at Murrumbateman, New South Wales. *Australian Journal of Soil Research* 24:49–60.

Aujla, T. S., and S. S. Cheema. 1983. Modifying profile water storage through tillage, herbicide, chemical evaporation retardant, and straw mulch and its effect on rainfed chick pea (*Cicer arietinum* L.). *Soil and Tillage Research* 3:159–170.

Bansal, S. P., P. R. Gajri, and S. S. Prihar. 1971. Effects of mulches on hydrothermal regime of soil and growth of maize and bajra. *Indian Journal of Agricultural Sciences* 41:467–473.

Black, A. L., and F. A. Siddoway. 1979. Influence of tillage and wheat straw residue management on soil properties in the Great Plains. *Journal of Soil and Water Conservation* 34:220–223.

Bond, J. J., and W. O. Willis. 1969. Soil water evaporation: Surface residue rates and placement effects. *Soil Science Society of America Proceedings* 33:445–448.

Borst, H. L., and H. J. Mederski. 1957. *Surface Mulches and Mulch Tillage for Corn Production.* Ohio Agricultural Experiment Station Research Bulletin no. 1796. Wooster, Ohio.

Carter, M. R., and D. A. Rennie. 1985. Soil temperature under zero tillage systems for wheat in Saskatchewan. *Canadian Journal of Soil Science* 65:329–338.

al-Darby, A. M., M. A. Mustafa, A. M. al-Omran, and M. O. Mohjoub. 1989. Effect of wheat residue and evaporative demands on intermittent evaporation. *Soil and Tillage Research* 15:105–116.

Enz, J. W., L. J. Brun, and J. K. Larson. 1988. Evaporation and energy balance for bare and stubble covered soil. *Agriculture and Forest Meteorology* 43:59–70.

Greb, B. W. 1967. Percent soil cover by six vegetative mulches. *Agronomy Journal* 5:610–611.

———. 1983. Water conservation: Central Great Plains. In *Dryland Agriculture,* edited by H. E. Dregne and W. O. Willis, pp. 57–72. Agronomy Monograph no. 23. Madison, Wis.: American Society of Agronomy.

Greb, B. W., D. E. Smika, and A. L. Black. 1967. Effect of straw mulch rates on soil water storage during summer fallow in the Great Plains. *Soil Science Society of America Proceedings* 31:556–559.

———. 1970. Water conservation with stubble mulch fallow. *Journal of Soil and Water Conservation* 25:58–62.

Gregory, J. M. 1982. Soil cover prediction with various amounts and types of crop residue. *Transactions of the American Society of Agricultural Engineers* 25:1333–1337.

Gupta, J. P., and G. N. Gupta. 1983. Effect of grass mulching on growth and yield of legumes. *Agricultural Water Management* 6:375–384.

Hanks, R. J., S. A. Bowers, and L. D. Bark. 1961. Influence of soil surface conditions on net radiation, soil temperature, and evaporation. *Soil Science* 91:233–238.

Hanks, R. J., and N. P. Woodruff. 1958. Influence of wind on vapor transfer through soil, gravel, and straw mulches. *Soil Science* 86:160–164.

Hundal, S. S., and S. K. De Datta. 1982. Effect of dry season soil management on water conservation for the succeeding rice crop in a tropical soil. *Soil Science Society of America Journal* 46:1081–1086.

Jacks, G. V., W. D. Brind, and R. Smith. 1955. *Mulching.* Bureau of Soil Technical Communication no. 49. Farnham Royal, Bucks, England: Commonwealth Agricultural Bureaux.

Jalota, S. K. 1990. Post-wetting soil water conservation with tillage and crop residue. In *Abstracts of the International Symposium on Water Erosion, Sedimentation, and Resource Conservation, Dehradun, India,* p. 68. New Delhi: Central Board of Irrigation and Power.

Jalota, S. K. and S. S. Prihar. 1979. Soil water storage and weed growth as affected by shallow tillage and straw mulching with and without herbicide in bare-fallow. *Indian Journal of Ecology* 5:41–48.

_____. 1990. Effect of straw mulch on evaporation reduction in relation to rates of mulching and evaporativity. *Journal of the Indian Society of Soil Science* 38:728–730.

Khera, K. L., R. Khera, S. S. Prihar, B. S. Sandhu, and K. S. Sandhu. 1976. Mulch, nitrogen, and irrigation effects on growth, yield, and nutrient uptake of forage corn. *Agronomy Journal* 68:937–941.

Lal, R. 1974. Effects of constant and fluctuating soil temperature on growth, development, and nutrient uptake of maize seedlings. *Plant and Soil* 40:589–606.

Lal, R., D. DeVleeschauwer, and R. M. Nganje. 1980. Changes in properties of a newly cleaned tropical Alfisol as affected by mulching. *Soil Science Society of America Journal* 44:827–833.

Lehane, I. 1961. Mulching of arable soils with organic material. *Soil and Fertilizer* 25:135.

Lemon, E. R. 1956. Potentialities for decreasing soil water loss. *Soil Science Society of America Proceedings* 20:160–163.

McCalla, T. M. 1947. Light reflection from stubble mulch. *Journal of the American Society of Agronomy* 39:690–691.

McCalla, T. M., and F. L. Duley. 1946. Effect of crop residues on soil temperature. *Journal of the American Society of Agronomy* 38:75–89.

Minhas, P. S., and A. S. Gill. 1985. Evaporation from soil as affected by incorporation and surface and sub-surface placement of oat straw. *Journal of the Indian Society of Soil Science* 33:774–778.

Nyborg, M., and S. S. Malhi. 1989. Effect of zero and conventional tillage on barley yield and nitrate nitrogen content, moisture, and temperature of soil in north-central Alberta. *Soil and Tillage Research* 15:1–19.

Ojeniyi, S. O. 1986. Effect of zero-tillage and disc ploughing on soil water, soil temperature, and growth and yield of maize (*Zea mays* L.). *Soil and Tillage Research* 7:173–182.

Prihar, S. S., S. K. Jalota, and J. L. Steiner. 1996. Residue management for evaporation reduction in relation to soil type and evaporativity. *Soil Use and Management* 12:150–157.

Prihar, S. S., B. Singh, and B. S. Sandhu. 1968. Influence of soil and environment on evaporation losses from mulched or unmulched pots. *Journal of Research, Punjab Agricultural University, Ludhiana, India,* 5:320–328.

Prihar, S. S., R. Singh, N. Singh, and K. S. Sandhu. 1979. Yield and water use of dryland crops as affected by mulching in standing maize and fallow. *Experimental Agriculture* 15:129–134.

Rasmussen, V. P., R. L. Newhall, and R. L. Cartee. 1986. Dryland conservation tillage systems. *Utah Science* 47:46–51.

Russel, J. C. 1939. The effect of surface cover on soil moisture losses by evaporation. *Soil Science Society of America Proceedings* 4:65–70.

Sandhu, B. S., K. L. Khera, and C. B. Singh. 1989. Straw mulch effects on evaporation, soil temperature, and crop growth. In *Proceedings of International Workshop on Evaporation from Open Water Surface (Vadodara, India),* pp. 155–159. New Delhi: Central Board of Irrigation and Power.

Saxton, K. E., K. L. Bristow, G. N. Flerchinger, and M. A. Omer. 1988. Tillage and Crop Residue Management. In *Proceedings of the International Conference "Dry Land Farming: A Global Perspective,"* edited by P. W. Unger, T. V. Sneed, W. R. Jorden, and R. L. Jensen, pp. 493–498. Amarillo/Bushland: Texas Agricultural Experiment Station.

Sharma, P. K., P. C. Kharwara, and R. K. Tewatia. 1990. Residual soil moisture and wheat yield in relation to mulching and tillage during preceding rainfed crop. *Soil and Tillage Research* 15:279–284.

Sharma, R. A., and R. K. Gupta. 1987. Soil water evaporation from clay and sandy loam soils as influenced by straw and dust mulching. *Indian Journal of Soil Conservation* 12:49–57.

_____. 1990. Conservation of soil moisture due to crop residue management and cultural practices for rainfed agriculture. In *Proceedings of International Symposium on Water Erosion, Sedimentation, and Resource Conservation (Dehradun, India),* pp. 233–243. New Delhi: Central Board of Irrigation and Power.

Sharma, R. A., M. S. Upadhay, and R. S. Tomar. 1985. Water use efficiency of some rainfed crops on a Vertisol as influenced by soil and straw mulching. *Journal of the Indian Society of Soil Science* 33:387–391.

Sharma, R. A., G. P. Verma, R. K. Gupta, and R. K. Katre. 1982. Moisture depletion pattern and use by unirrigated wheat and safflower grown on Vertisol as influenced by cultural practices. *Zeitschrift für Acker und Pflazenbau* 151:257–264.

Sharma, R. A., G. P. Verma, and R. K. Gupta. 1985. Modification of evaporation from a Vertisol by straw mulch. *Journal of the Indian Society of Soil Science* 33:383–386.

Singh, B., and B. S. Sandhu. 1979. Effect of irrigation, mulch, and crop canopy on soil temperature in forage maize. *Journal of the Indian Society of Soil Science* 27:225–236.

Singh, D., and S. D. Nijhawan. 1944. A critical study of the effect of soil mulch on evaporation of soil moisture. *Indian Journal of Agricultural Sciences* 14:127–137.

Smika, D. E. 1983. Soil water change as related to position of straw mulch on the soil surface. *Soil Science Society of America Journal* 47:988–991.

Soleneker, L. L., and W. C. Moldenhauer. 1977. Measuring the amounts of crop residue remaining after tillage. *Journal of Soil and Water Conservation* 32:231–236.

Steiner, J. L. 1989. Tillage and surface residue effects on evaporation from soils. *Soil Science Society of America Journal* 53:911–916.

Steiner, J. L., S. K. Jalota, and S. S. Prihar. 1990. Evaporation from soil: Texture, evaporativity, residue, and tillage effects. In *Agronomy Abstracts*, p. 23. Madison, Wis.: American Society of Agronomy.

Turkey, R. B., and E. L. Schoff. 1963. Influence of different mulching materials upon the soil environment. *American Society of Horticulture Science Proceedings* 82:68–76.

Unger, P. W. 1978. Straw-mulch effects on soil temperature and sorghum germination and growth. *Agronomy Journal* 70:858–864.

————. 1984. *Tillage Systems for Soil and Water Conservation*. FAO Soils Bulletin no. 54. Rome.

————. 1986. Wheat residue management effects on soil water storage and corn production. *Soil Science Society of America Journal* 50:764–770.

Unger, P. W., and J. J. Parker. 1968. Residue placement effect on decomposition, evaporation, and soil moisture distribution. *Agronomy Journal* 60:469–472.

————. 1976. Evaporation reduction from soil with wheat, sorghum, and cotton residues. *Soil Science Society of America Journal* 40:938–942.

Urmani, N. K., K. S. Pharande, and S. Quamaurzzman. 1973. Mulching conserves extra moisture. *Indian Farming* 23:24–26.

van Doren, D. M., Jr., and R. R. Allmaras. 1978. Effect of residue management practices on the soil physical environment, microclimate, and plant growth. In *Crop Residue Management Systems*, edited by W. R. Oschwald, pp. 49–83. American Society of Agronomy Special Publication 31. Madison, Wis.

Wall, D. A., and E. H. Stobbe. 1984. The effect of tillage on soil temperature and corn (*Zea mays* L.) growth in Manitoba. *Canadian Journal of Plant Science* 64:59–67.

Willis, W. O. 1962. Effect of partial covers on evaporation from soil. *Soil Science Society of America Proceedings* 90:598–601.

Wind, G. P. 1961. *Capillary Rise and Some Applications of the Theory of Moisture in Unsaturated Soil*. Institute of Land and Water Management Research Technical Bulletin 22. Wageningon, the Netherlands.

Wischmeier, W. H., and D. D. Smith. 1978. *Predicting Rainfall Erosion Losses—A Guide to Conservation Planning*. Agricultural Handbook no. 537. Washington, D.C.: U.S. Department of Agriculture.

6

Evaporation Reduction by
Combining Mulching and Tillage

Loosening the soil surface by tillage changes the water retention and transmission properties of soil. The hydraulic conductivity of the tilled layer vis-à-vis that of the untilled soil is greatly reduced, which retards the upward liquid flux across the tilled layer. With this restricted upward supply of water, a dry surface layer develops more rapidly, which lowers the locus of phase change to some distance below the surface and acts as a barrier to the escape of vapor to the atmosphere. Since the soil under residue mulch remains wet for a longer period, upward liquid flow of water also continues for longer periods. In some cases, cumulative water loss from a residue-covered soil may exceed that from bare soil in a prolonged drying cycle (Jacks, Brind, and Smith, 1955; McCalla and Army, 1961). These observations lead one to postulate that the benefit of evaporation reduction with residue in such situations could be prolonged if the upward water supply under the residue were cut off by disrupting the capillary flow (Brun et al., 1986). In other words, a combination of residue cover and tillage could bring about greater evaporation reduction than residue cover alone. Experiments to verify these suppositions are discussed in this chapter.

In the previous chapter we described results that revealed a decline in evaporation reduction with straw mulch after a peak had been attained. Recent studies by Prihar, Jalota, and Steiner (1996), however, have shown that when either the straw was mixed with surface soil by tillage or the soil was undercut below the residue kept at the soil surface, the decline in evaporation reduction after the peak was retarded and checked. However, the length of time peak evaporation reduction was maintained, and hence the longevity of the benefit, depended upon soil type, evaporativity (E_0), and mulch rate used.

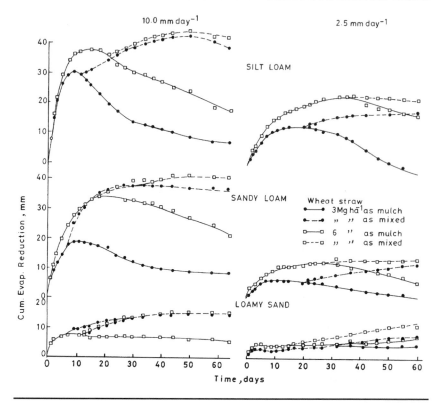

FIGURE 6.1. Time course of cumulative evaporation reduction in different soils under 3 and 6 Mg ha⁻¹ wheat straw applied as mulch or mixed with the top 5 cm of soil under evaporativities of 2.5 and 10.0 mm day⁻¹ (Prihar, Jalota, and Steiner, 1996)

Prihar, Jalota, and Steiner (1996) studied the effect of differently managed 3 and 6 Mg ha⁻¹ wheat (*Triticum aestivum* L.) straw mulches on evaporation reduction in 95 cm long and 10 cm internal diameter columns of silt loam, sandy loam, and loamy sand soils under evaporativities of 2.5 and 10 mm day⁻¹. The residue management treatments were as follows: 5 cm deep undercut with chopped straw at the surface; chopped straw mixed in the surface 5 cm of soil; and 5 cm deep tillage after removal of straw. All treatments were given as soon as the rate of loss from mulched columns exceeded that from the unmulched columns. Residue mixed by shallow tillage in the top 5 cm of the soil changed the course of cumulative evaporation reduction (CER). In a drying cycle of 60 days, the maximum evaporation reduction (MER) and the time of its occurrence varied with E_0, residue rate, and soil texture (Figure 6.1).

TABLE 6.1 Magnitude and longevity of maximum evaporation reduction (MER) as influenced by wheat straw management and soil type (E_0 = 10 mm day^{-1})

Treatment	3 Mg ha^{-1}		6 Mg ha^{-1}	
	MER (mm)	No. of days MER sustained	MER (mm)	No. of days MER sustained
		Silt loam		
Residue spread as mulch	30	9	38	15
Soil tilled after removing the residue	28	40	34	44
Soil tilled with residue at the surface	26	40	33	40
Residue mixed with top 0–5 cm	43	50	44	50
		Sandy loam		
Residue spread as mulch	18	9	34	18
Soil tilled after removing the residue	25	18	29	30
Soil tilled with residue at the surface	31	34	41	44
Residue mixed with top 0–5 cm	37	50	41	50

Source: Prihar, Jalota, and Steiner, 1996.

Effect of Soil Type

In the above study by Prihar, Jalota, and Steiner (1996), CER with 3 Mg ha^{-1} mulch under E_0 of 10.0 mm day^{-1} peaked at 30 mm in silt loam and 18 mm in sandy loam at 9 days in both the soils. With 6 Mg ha^{-1} mulch, CER peaked at 38 mm at 15 days in the silt loam and 34 mm at 18 days in the sandy loam. As already discussed, the loosening of the surface layer decreased the hydraulic conductivity and retarded the upward liquid flow (Jalota and Prihar, 1992). This effect was accentuated when straw was mixed in the tilled zone because straw reduced the effective cross-sectional area contributing to liquid flow. These changes brought the rate of evaporation from residue-mixed pots down almost to that from the unmulched pots. The decline in the peak CER was stalled, and the maximum and longest-lasting benefit occurred in the silt loam, followed by the sandy loam soil. The effect on the course of evaporation reduction by tillage after removal of residue and undercut was similar to that of the residue mixed. But the MER was generally lower and was sustained for shorter periods with tillage and undercut than with residue mixed (Table 6.1). Moreover, the MER and its longevity were greater with 6 Mg ha^{-1} than with 3 Mg ha^{-1} for mulching and undercut treatments, but not for the residue mixed. It appears that for both soils 3 Mg ha^{-1} straw is sufficient to derive maximum benefit when mixed with the surface 5 cm.

TABLE 6.2   Maximum evaporation reduction (MER) and its time of occurrence as influenced by straw mulch rates and mixing depths on silty clay loam (Sicl) and sandy loam (Sl) soils under different evaporativities

Treatments		MER (mm)			
		$E_0 = 8.7$ mm day^{-1}		$E_0 = 2.0$ mm day^{-1}	
Mulch rate (Mg ha^{-1})	Mixing depth (cm)	Sicl	Sl	Sicl	Sl
0	2	19 (21)	17 (19)	46 (82)	38 (78)
0	5	45 (46)	39 (46)	51 (91)	47 (91)
2	0	28 (15)	21 (11)	35 (70)	27 (68)
2	2	33 (29)	23 (17)	51 (91)	52 (91)
2	5	47 (46)	41 (46)	55 (91)	60 (91)
4	0	49 (17)	31 (14)	54 (84)	45 (78)
4	2	55 (30)	37 (23)	53 (91)	46 (91)
4	5	64 (46)	46 (46)	53 (91)	46 (91)
8	0	67 (23)	43 (16)	70 (91)	62 (91)
8	2	80 (37)	44 (37)	—	—
8	5	75 (46)	55 (46)	—	—

Source: Gill and Jalota, 1996.
Note: Figures in parentheses indicate the time in days at which MER occurred.

Effect of Depth of Mixing

Gill and Jalota (1996) reported that apart from soil type, E_0, and residue rate, the magnitude and time of occurrence of MER were also influenced by the depth of mixing of residue (Table 6.2). For example, under an E_0 of 8.7 mm day^{-1}, MER for 2, 4, and 8 Mg ha^{-1} of rice (*Oryza sativa* L.) residue mixed with the upper 2 cm layer was 33, 55, and 80 mm at 29, 30, and 37 days in silty clay loam and 23, 37, and 44 mm at 17, 23, and 37 days in sandy loam soil, respectively. But when the same rates were mixed with the top 5 cm, the MER increased to 47, 64, and 75 mm in silty clay loam and to 41, 46, and 55 mm in sandy loam at the end of the study (46 days). More interestingly, after a certain period CER with a smaller amount of straw mixed with the top 0–5 cm of soil exceeded that with a higher amount used as mulch or mixed with the top 0–2 cm layer for both soils and both evaporativities. For example, in the sandy loam soil CER with 2 Mg ha^{-1} residue mixed to 5 cm depth exceeded that with 4 Mg ha^{-1} and 8 Mg ha^{-1} spread as mulch at 19 and 32 days, respectively (Figure 6.2b). Similar trends were obtained in the silty clay loam (Figure 6.2a). Significantly, the CER with 2 Mg ha^{-1} residue mixed with the 0–5 cm soil layer exceeded that with 4 Mg ha^{-1} mixed with the 0–2 cm soil layer at 45 days in the silty clay loam and 42 days in the sandy loam. In the sandy loam CER with 2 Mg ha^{-1} residue mixed with the 0–5 cm layer equaled that with 8 Mg ha^{-1} mixed with the 0–2 cm layer at 46 days.

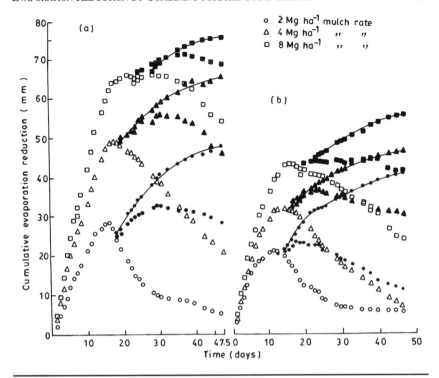

FIGURE 6.2. Time course of cumulative evaporation reduction from *(a)* silty clay loam and *(b)* sandy loam soils as affected by straw mulch rates alone *(hollow points)* and by mixing with top 2 cm *(solid points)* and top 5 cm *(continuous line and solid points)* of the soil under E_0 of 8.7 mm day^{-1} (Gill and Jalota, 1996)

Gill (1992) studied under field conditions the effect of the amount and depth of mixing of rice straw on evaporation for 43 days (from 275 to 318 Julian days). Rice straw was added as mulch or mixed with the 2 cm or the 5 cm surface layer at the rate of 0, 2, 4, and 8 Mg ha^{-1}. Evaporation was computed by subtracting the amount of drainage from the total change in soil water storage at 2-day intervals. Drainage was computed by plugging the predetermined values of the constants A and B into Ogata and Richards's (1957) equation:

$$\frac{dw}{dt} = AB(W/A)^{B-1/B}$$

where dw/dt is change in soil water storage (W) (cm) in the 0–100 cm soil layer at time t (days). Cumulative evaporation loss during 43 days decreased while the drainage component increased with increasing mulch rate (Table 6.3). Mixing the straw in the 0–5 cm layer was most effective in reducing evaporation losses. Mixing to only 2 cm deep was less effective.

TABLE 6.3   Evaporation (E) and drainage (D) components (in mm) from 100 cm deep soil profiles at 43 days as affected by paddy straw mulch rates and mixing depths

Mixing depth (cm)	0 Mg ha⁻¹		2 Mg ha⁻¹		4 Mg ha⁻¹		8 Mg ha⁻¹	
	E	D	E	D	E	D	E	D
0	60.8	5.7	53.4	16.0	45.9	18.0	29.2	31.2
2	58.1	6.0	51.0	18.0	42.7	19.5	25.1	34.7
5	54.8	7.9	48.7	18.8	38.8	19.8	22.0	38.8
Mean	57.9	6.5	51.0	17.6	42.5	19.1	25.7	34.1
LSD (0.05) for E								
Mulch	5.0							
Mixing depth	4.4							
Interaction	NS							

Source: Gill, 1992.
Note: LSD = least significant difference. NS = not significant.

Effect of Rainfall

Papendick and Parr (1987) simulated the effects of no-till and tillage with and without crop residue on evaporation from a deep loam soil. Tillage in the residue-covered soil had little effect on evaporation reduction, especially under frequent wetting. Although initially evaporation from the tilled bare soil was higher than from the residue-covered soil, the cumulative evaporation at the end of the simulation run (10 weeks) was the same in the two cases. In the absence of frequent precipitation, the initial evaporation rate and cumulative loss were highest in the bare untilled soil. At the end of simulation, cumulative evaporation from no-till residue-covered soil approached that from bare untilled soil. But the cumulative evaporation from the tilled residue-covered soil was half (or less) of that from the untilled soil (Figure 6.3). The cumulative evaporation from the tilled residue-covered soil exceeded that from the tilled bare soil.

Jalota (1990) studied the effect of timing of tillage and mulch application on evaporation reduction. In the absence of tillage, 3 Mg ha⁻¹ wheat straw mulch applied 5 days after wetting to silt loam soil columns exposed to an E_0 of 6.3 mm day⁻¹ reduced cumulative evaporation only slightly (Figure 6.4). Reduction with 6 Mg ha⁻¹ was greater than with 3 Mg ha⁻¹. However, when mulch was applied after 5 cm deep tillage, evaporation with both mulch rates was almost equal to that under tillage without mulch. This shows that residue applied after tillage was of little use. Yet the evaporation with these treatments was substantially lower than that from the untilled columns.

Rydberg (1990) reported that "ploughless tillage" similar to stubble mulch that left crop residue at the surface and incorporated residue reduced evapora-

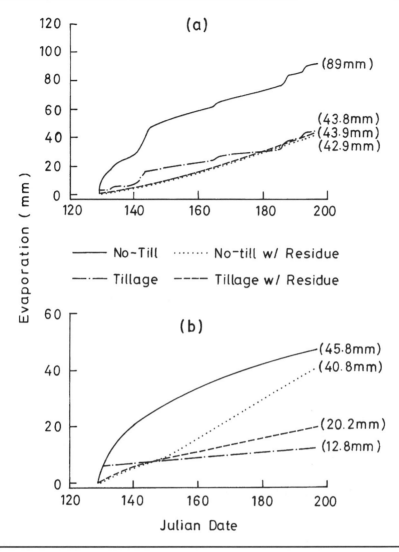

FIGURE 6.3.   Simulated cumulative evaporation as affected by tillage and residue for an initially moist loam soil *(a)* with and *(b)* without precipitation, using weather inputs for 1986 for Pullman, Washington (the number after each curve is the total evaporation at the end of simulation) (Papendick and Parr, 1987)

tion relative to that when all residues were removed after harvest. In contrast, al-Darby et al. (1989) reported no effect on evaporation from incorporated residue in laboratory studies on soil drying. In fact, the impact of incorporated residue on evaporation depends upon the changes in hydraulic conductivity of the soil as it dries. In many soils incorporation of residue may reduce hydraulic

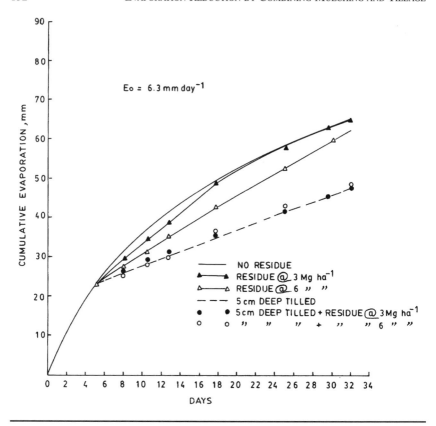

FIGURE 6.4. Cumulative evaporation from silt loam soil as influenced by mulch rates and tillage under an E_0 of 6.3 mm day^{-1}. Treatments were given 5 days after wetting.

conductivity by destroying pore geometry. Baumhardt, Zartman, and Unger (1985) evaluated the effect of disk and no-tillage management of wheat residue on soil water storage on Pullman clay loam at Bushland and Lubbock, Texas. Soil water storage was greater with no tillage than with disking at Bushland, where wheat produced 11 Mg ha^{-1} of residue. At Lubbock, where the residue amount was only 2 Mg ha^{-1}, water storage differences were slight. Smika (1976) summarized the effect of clean tillage and stubble mulching on soil water gain at several locations in the United States (Table 6.4). At wheat planting the average water storage was 27 mm greater with stubble mulch than with clean tillage. Some recent studies (Laryea and Unger, 1995; Unger, 1993; unpublished data of Ordie R. Jones) have shown that no tillage conserved more moisture than sweep and moldboard plow tillage on Pullman clay loam at Bushland. The authors attributed this to the fact that no-till soil develops cracks and facilitates deep penetration of rain water, where it is less susceptible to evaporation.

TABLE 6.4 Net gain in soil water during fallow with clean and stubble mulch tillage at seven central Great Plains locations (USA)

Location	Years of data	Net gain in soil water (mm)	
		Clean	Stubble mulch
Akron, Colo.	6	142	173
Colby, Kans.	4	115	141
Garden City, Kans.	6	86	90
Oakley, Kans.	4	82	131
North Platte, Nebr.	8	146	203
Alliance, Nebr.	8	29	32
Archer, Wyo.	2	28	42
Weighted average		95	122

Source: Smika, 1976.

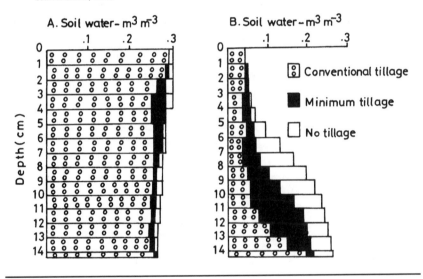

FIGURE 6.5.   Soil water content to a 15 cm depth *(A)* 1 day and *(B)* 34 days after 165 mm of rainfall as influenced by tillage treatments (Smika, 1976)

As discussed earlier, the amount and position of residue at the surface strongly affect the initial evaporation rates from soil. Once the soil surface dries out, the rate of upward water flow and the porosity, or air permeability, of the surface soil become more important in determining the evaporation rate. This was illustrated by Smika (1976), who compared the effects of conventional, minimum, and no-tillage treatments on soil water loss during a 34-day rainless period following 165 mm of rain. On the day after rain, soil water contents in the 0–15 cm layer were comparable in all treatments (Figure 6.5A). At the end of 34 days, soil receiving the conventional tillage had dried to less than 0.1 m^3 m^{-3} to

TABLE 6.5   Effect of tillage methods on average
precipitation storage during an 11-month fallow
period in irrigated wheat–dryland sorghum
cropping system on Pullman clay loam in Texas

Tillage method	Precipitation storage (%)
No tillage	35
Sweep	23
Disk	15

Source: Unger and Wiese, 1979.
Note: Precipitation was 348 mm.

12 cm depth, and soil under minimum tillage had dried to 9 cm depth (Figure 6.5B). These depths corresponded to the depths at which the blade tillage operations were performed 8 days before rain. In contrast, untilled soil dried to less than 0.1 m^3 m^{-3} water content only to a 5 cm depth. Some soil water loss occurred from depths greater than these under all treatments, but the soil water content was highest under no-tillage at all the depths. Amounts of residue left at the surface during the 34-day drying cycle were 1.20, 2.20, and 2.70 Mg ha^{-1} with conventional, minimum, and no tillage, respectively.

At Bushland, Unger and Wiese (1979) used no tillage, sweep tillage, and disk tillage for residue management and weed control during an 11-month period following harvest of irrigated winter wheat until planting of dryland sorghum. Precipitation storage during the fallow was highest with no tillage and lowest with disk tillage (Table 6.5).

Effect of Crop-Residue Mulching and Mixing on the Moisture Profile

Like bare soil, soil mulched with crop residues also exhibits continuous soil water content profiles with depth. However, the moisture content gradients are lower in mulched than in unmulched soil. When residue is mixed into the surface layer, the surface layer dries out faster than the corresponding depth in untilled and residue-covered soil for the reasons discussed in Chapter 2. The phenomenon is more pronounced under higher E_0 and lower residue rate and in coarser-textured soils. Gill and Jalota (1996) observed that the postdrying soil water content of a 2.5–3.75 cm soil layer was higher when 4 Mg ha^{-1} wheat residue was mixed with the top 2 cm than when it was mixed with the top 5 cm or used as mulch (Figure 6.6). In silty clay loam, the water content of the 2.5–3.75 cm layer after 97 days of drying under an E_0 of 2.5 mm day^{-1} averaged 30.4% (by volume) for

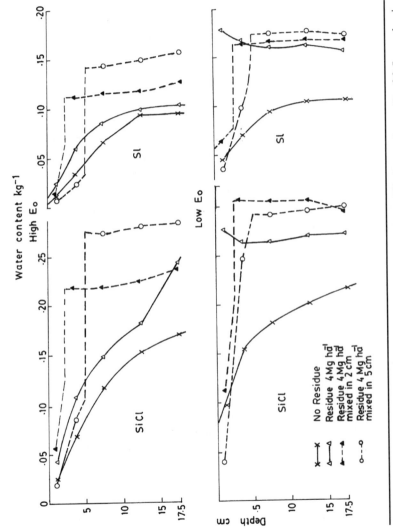

FIGURE 6.6. Near-surface soil moisture profiles after 97 days of drying under an E_0 of 2.5 mm day[-1] and after 46 days under an E_0 of 8.7 mm day[-1] as affected by residue rate, mixing depth, soil type, and E_0 (Gill and Jalota, 1996)

2 cm deep residue mixing and 17% for 5 cm deep mixing. This was true for other soils and evaporativities too.

In the sandy loam, the corresponding moisture contents of this layer were 16.4% and 8.2% for 2 cm and 5 cm deep mixing. Under an E_0 of 8.7 mm day^{-1}, the soil water contents of this layer at 46 days of drying were 16.5% and 9.0% higher with 2 cm deep residue mixing than with 5 cm deep mixing in silty clay loam and sandy loam soils, respectively. Thus, shallow cultivation of soil with or without crop residues causes higher wetness in layers immediately below the depth of tillage or residue mixing and offers potentialities for the retention of moisture in the seed zone over long dry periods.

Summary

Evaporation reduction with straw mulching is a time-variant phenomenon (see Chapter 5). It increases up to a certain period, attains a peak (maximum evaporation reduction), and declines thereafter. The decline in cumulative evaporation reduction is caused by a reversal of the evaporation rates from bare and mulched soils; the former is initially higher. The reversal occurs because the surface layer under mulch remains wetter than bare soil, which dries out. This decline in peak reduction can be offset by reducing the rate of replenishment of water from lower layers to the surface in mulched soil either by interfering with the capillarity of the soil or by mixing the residue into the top few centimeters of the soil by tillage. Mixing in residue was found to be more effective because it decreased the cross-sectional area contributing to liquid flow in addition to destroying capillarity. Therefore, a combination of residue and tillage was found to cause greater evaporation reduction than tillage or residue alone. The effect of tillage in combination with residue was greater in finer-textured soils and under high evaporativity. The effect of residue spread as mulch after tillage was negligible. Therefore, the combination of keeping the soil surface covered with crop residue during the initial period of drying and of mixing the residue with a few centimeters of surface soil when the evaporation rate from mulched soil exceeds that from bare soil holds promise for evaporation reduction.

References

Baumhardt, R. L., R. E. Zartman, and P. W. Unger. 1985. Grain sorghum response to tillage methods used during fallow and to limited irrigation. *Agronomy Journal* 77:643–646.

Brun, L. J., J. W. Enz, J. K. Larsen, and C. Fanning. 1986. Springtime evaporation from bare and stubble-covered soil. *Journal of Soil and Water Conservation* 41:120–122.

al-Darby, A. M., M. A. Mustafa, A. M. al-Omran, and M. O. Mohjoub. 1989. Effect of wheat residue and evaporative demands on intermittent evaporation. *Soil and Tillage Research* 15:105–116.

Gill, B. S. 1992. Interactive effects of paddy straw mulch rates and tillage depths on evaporation reduction in relation to soil type and atmospheric evaporativity. M.Sc. thesis, Punjab Agricultural University, Ludhiana, India.

Gill, B. S., and S. K. Jalota. 1996. Evaporation from soil in relation to residue rate, mixing depth, soil texture, and evaporativity. *Soil Technology* 8:293–301.

Jacks, G. V., W. D. Brind, and R. Smith. 1955. *Mulching*. Bureau of Soil Technical Communication no. 49. Farnham Royal, Bucks, England: Commonwealth Agricultural Bureaux.

Jalota, S. K. 1990. Post-wetting soil water conservation with tillage and crop residue. In *Abstracts of the International Symposium on Water Erosion, Sedimentation, and Resource Conservation, Dehradun, India,* p. 68. New Delhi: Central Board of Irrigation and Power.

Jalota, S. K., and S. S. Prihar. 1992. Liquid component of evaporative flow in two tilled soils. *Soil Science Society of America Journal* 56:1893–1898.

Laryea, K. B., and P. W. Unger. 1995. Grassland converted to crop land: Soil conditions and sorghum yields. *Soil and Tillage Research* 33:29–45.

McCalla, T. M., and T. J. Army. 1961. Stubble mulch farming. *Advances in Agronomy* 13:125–196.

Ogata, G., and L. A. Richards. 1957. Soil water content changes following irrigation of bare-field soil that is protected from evaporation. *Soil Science Society of America Proceedings* 21:355–356.

Papendick, R. I., and J. F. Parr. 1987. Soil, crop, and water management systems for rainfed agriculture. In *Proceedings of an International Workshop on Soil, Crop, and Water Management* (ICRISAT Sahelian Centre, Niamey, Niger). Patancheru, Andhra Pradesh, India: International Crops Research Institute for Semi-arid Tropics.

Prihar, S. S., S. K. Jalota, and J. L. Steiner. 1996. Residue management for evaporation reduction in relation to soil type and evaporativity. *Soil Use and Management* 12:150–157.

Rydberg, T. 1990. Effects of ploughless tillage and straw incorporation on evaporation. *Soil and Tillage Research* 17:303–314.

Smika, D. E. 1976. Seed zone soil water conditions with reduced tillage in the semi-arid central Great Plains. In *Proceedings of the 7th Conference of the International Soil Tillage Research Organization,* pp. 37.1–37.6. Uppsala, Sweden: International Soil Tillage Research Organization.

Unger, P. W. 1993. Tillage effects on dryland wheat and sorghum production in southern Great Plains. *Agronomy Journal* 86:310–314.

Unger, P. W., and A. F. Wiese. 1979. Managing irrigated winter wheat residues for water storage and subsequent dryland sorghum production. *Soil Science Society of America Journal* 43:582–588.

Index

139

ISBN 0-8138-2857-0

90000